DEFENSE, SECURITY AND STRATEGIES

CHALLENGES TO THE DEFENSE INDUSTRIAL BASE

DEFENSE, SECURITY AND STRATEGIES

Additional books in this series can be found on Nova's website under the Series tab.

Additional E-books in this series can be found on Nova's website under the E-book tab.

DEFENSE, SECURITY AND STRATEGIES

CHALLENGES TO THE DEFENSE INDUSTRIAL BASE

LEENA NWOSU
EDITOR

nova publishers
New York

For permission to use material from this book please contact us:
Telephone 631-231-7269; Fax 631-231-8175
Web Site: http://www.novapublishers.com

Additional color graphics may be available in the e-book version of this book.

Library of Congress Cataloging-in-Publication Data

ISBN: 978-1-62618-141-0

Published by Nova Science Publishers, Inc. † New York

CONTENTS

PREFACE

The Panel on Business Challenges within the Defense Industry conducted hearings, held roundtable discussions with industry, and reviewed many studies and publications that have examined the Defense Industrial Base (DIB) in order to gain insight into the complexity of the industrial base and the wide variety of factors influencing business behavior and the shaping of the DIB. These efforts highlighted that the DIB is unique in that the Department of Defense (DOD) plays a role as both the customer and regulator of the DIB. This book provides several recommendations that the Panel believes will improve the defense business environment, reduce barriers to entry, spur innovation, increase competition, and aid in getting critical technology into the hands of the Warfighter.

Chapter 1 - The House Committee on Armed Services (HASC) Panel on Business Challenges within the Defense Industry was appointed by Chairman Howard P. "Buck" McKeon and Ranking Member Adam Smith on September 12, 2011 to examine the challenges that businesses face when working with the Department of Defense (DOD). Chairman McKeon and Ranking Member Smith recognized the need for the HASC to form a panel to spend six months focusing on this specific issue.

Under Rule 5(a) of the rules of the Committee, seven Committee members were asked to examine several specific areas in addition to their normal responsibilities on the Committee. Appointed to the Panel were Congressman Bill Shuster, Congressman Rick Larsen, Congressman Bobby Schilling, Congresswoman Betty Sutton, Congressman Jon Runyan, Congresswoman Colleen Hanabusa, and Congressman Allen West.

The Panel was asked to examine (1) contracting or regulatory challenges facing the defense industry; (2) the use of incentives and mandates to shape the defense industrial base; (3) structural challenges facing various sectors within the industrial base, including universities and research institutes; (4) the impact of the current fiscal environment on the health of the defense industrial base at both the prime and subcontractor levels; and (5) opportunities to reduce barriers to entry. With these criteria, the Panel's findings and recommendations reported back to the full membership of the HASC for action.

In order to obtain all of the necessary information regarding these matters, the Panel maintained a collaborative effort with both members of industry and officials from the DOD. This effort took shape in a combination of formal hearings, with various witnesses from the government and private sector, as well as think tanks and associations. The Panel also took briefings to gather information on current practices within the DOD. In additional to the formal events held in Washington, DC, the Panel also traveled across the country to hold roundtable discussions with the defense industry, research organizations and the organic industrial base. These field events were incredibly informative and were a critical part of the Panel's efforts to examine the challenges facing businesses wanting to provide for and support the Warfighter.

Chapter 2 - DOD's need for TWVs dramatically increased in response to operational demands and threats experienced in Afghanistan and Iraq. TWVs primarily transport cargo and personnel in the field and include the High Mobility Multi-purpose Wheeled and Mine Resistant Ambush Protected vehicles. The U.S. TWV industrial base, which includes manufacturers and suppliers of major subsystems, increased production to meet DOD's wartime requirements. That base now faces uncertainties as DOD's budget declines and operational requirements for these vehicles decrease. In addition to sales to DOD, U.S. manufacturers sell vehicles to foreign governments.

The Senate Armed Services Committee Report on a bill for the National Defense Authorization Act for Fiscal Year 2012 directed GAO to (1) describe the composition of the U.S. TWV industrial base, (2) determine how many U.S. manufactured TWVs were purchased by foreign governments from fiscal years 2007 through 2011, and (3) identify factors perceived as affecting foreign governments' decisions to purchase these vehicles. GAO analyzed data from DOD on U.S. and foreign government TWV purchases, as well as sales data from the four primary U.S. TWV manufacturers. GAO also collected data from five foreign governments, including those that did and did not purchase U.S. TWVs.

In: Challenges to the Defense Industrial Base ISBN: 978-1-62618-141-0
Editor: Leena Nwosu

Chapter 1

CHALLENGES TO DOING BUSINESS WITH THE DEPARTMENT OF DEFENSE*

House Committee on Armed Services, Panel on Business Challenges in the Defense Industry

ABSTRACT

The House Committee on Armed Services (HASC) Panel on Business Challenges within the Defense Industry was appointed by Chairman Howard P. "Buck" McKeon and Ranking Member Adam Smith on September 12, 2011 to examine the challenges that businesses face when working with the Department of Defense (DOD). Chairman McKeon and Ranking Member Smith recognized the need for the HASC to form a panel to spend six months focusing on this specific issue.

Under Rule 5(a) of the rules of the Committee, seven Committee members were asked to examine several specific areas in addition to their normal responsibilities on the Committee. Appointed to the Panel were Congressman Bill Shuster, Congressman Rick Larsen, Congressman Bobby Schilling, Congresswoman Betty Sutton, Congressman Jon Runyan, Congresswoman Colleen Hanabusa, and Congressman Allen West.

The Panel was asked to examine (1) contracting or regulatory challenges facing the defense industry; (2) the use of incentives and mandates to shape the defense industrial base; (3) structural challenges

* This report was released by the House Committee on Armed Services on March 19, 2012.

facing various sectors within the industrial base, including universities and research institutes; (4) the impact of the current fiscal environment on the health of the defense industrial base at both the prime and subcontractor levels; and (5) opportunities to reduce barriers to entry. With these criteria, the Panel's findings and recommendations reported back to the full membership of the HASC for action.

In order to obtain all of the necessary information regarding these matters, the Panel maintained a collaborative effort with both members of industry and officials from the DOD. This effort took shape in a combination of formal hearings, with various witnesses from the government and private sector, as well as think tanks and associations. The Panel also took briefings to gather information on current practices within the DOD. In additional to the formal events held in Washington, DC, the Panel also traveled across the country to hold roundtable discussions with the defense industry, research organizations and the organic industrial base. These field events were incredibly informative and were a critical part of the Panel's efforts to examine the challenges facing businesses wanting to provide for and support the Warfighter.

EXECUTIVE SUMMARY

The Panel on Business Challenges within the Defense Industry conducted hearings, held roundtable discussions with industry, and reviewed many studies and publications that have examined the Defense Industrial Base (DIB) in order to gain insight into the complexity of the industrial base and the wide variety of factors influencing business behavior and the shaping of the DIB. These efforts highlighted that the DIB is unique in that the Department of Defense (DOD) plays a role as both the customer and regulator of the DIB.

These efforts also aided the Panel in identifying challenges to ensuring that the industrial base is positioned to support the needs of the nation in the 21st century. These challenges include the lack of a comprehensive DOD strategy for managing and maintaining an industrial base, the inconsistent communications that often marks the program office/private industry relationship, the fragmented nature of the industrial base.

The Panel found that small and midsize businesses face particular challenges in contracting with the DOD. In recent years, the DOD has been unable to meet its small business federal procurement goals. Also, because DOD's contracts tend to be fairly large, the Small Business Administration's limit of $2 million per contract for surety bonds is often not sufficient to be of use to small businesses contracting with the DOD. In addition, businesses that are other than small are excluded from set aside competitions under the authority of the Small Business Act. As a result, some small businesses that qualify for a set aside under the authority of the Act cannot obtain a required surety bond and some midsize businesses that can secure a required surety bond do not qualify for a set aside under the authority of the Act. Furthermore, the Panel found that DOD lacks the ability to track small business participation at the lower subcontract tiers. The Panel also found that DOD lacks a culture that fosters small business participation where appropriate.

In its work, the Panel concluded that the defense acquisition workforce has struggled to manage and execute programs in the midst of challenges in acquiring and retaining a professionally certified and competent defense acquisition workforce. Many described the defense acquisition workforce as part of a "risk-averse" culture where program managers, in particular, have difficulty in balancing risks in managing cost, schedule and performance of acquisition programs.

DOD has, in some cases, outsourced program management and divested itself of critical skills that are difficult to develop – contracting officials, cost estimators, and systems engineers. This reliance on private contractors can create a potential conflict of interest and blur the lines between what work must be performed by federal employees and what work is permitted to be performed by private contractors. The Panel notes that just as it takes many years to develop a military leader capable of commanding at the senior ranks of the operational force, it takes a similar amount of time to develop an acquisition professional with the knowledge, skills, and experience needed to manage large defense acquisition efforts. In addition, the Panel found that constantly changing regulations leads to unnecessary complexity, confusion, and poor execution, only furthering challenges for the acquisition workforce. The Panel also found that the DOD acquisition system lacks sufficient emphasis on small business participation.

Furthermore, the Panel found that DOD lacks a clearly articulated strategy that would provide a corporate vision of DOD's future technology needs. Starting in 1989, DOD was required to submit a Critical Technologies Plan that eventually morphed into the Defense Technology Area Plan and then supported by the Joint Warfighting Science and Technology Plan and the Defense Technology Objectives. Over the years, these requirements have been decreased resulting in a lack of strategy. This lack of strategy makes it difficult for industry to have visibility into the future developmental needs of the DOD and makes it challenging for both industry and the government to make investment decisions, such as those needed for internal IR&D and management of the supply chain. Furthermore, the Panel found that DOD acquisition policies and processes provide little incentive for defense industry to invest in innovation.

Additionally, the DOD has a multiplicity of uncoordinated funding mechanisms for technology transition to get early stage research and development efforts across the "valley of death." The GAO identified more than 20 initiatives for rapidly satisfying urgent operational needs from the Warfighter. Furthermore, there are a number of technology transition mechanisms such as Defense Production Act Title III, ManTech, Small Business Innovative Research program, and the Industrial Base Innovation Fund that have been implemented to serve slightly different but related purposes. Recent efforts to speed technology to the battlefield, such as the Rapid Innovation Fund, have also been added. The Government Accountability Office observed that the reach of these initiatives is limited and there is no unified, corporate approach to using them and noted that the Department's approach to funding transition is flawed and that multiple, small funding sources for specific transition activities offer a piecemeal solution to a more systemic problem.[1]

The Panel also found that a number of hurdles make it challenging for companies to compete for defense contracts. The plethora of regulations specific to government and defense contracting dissuades many companies from competing for government contracts. The acquisition process is often bureaucratic and rigid, with insufficient flexibility to allow appropriate application of management, oversight, and monitoring of small businesses. The defense business environment is also complicated, and some argue hindered, by current export control requirements. The high rate of personnel turnover in government acquisition personnel, from program managers to Defense Contract Audit Agency (DCAA) auditors affects the quality and consistency of policies. Oversight and management agencies such as DCAA are under-resourced and lack consistently trained, skilled personnel, hampering the ability of these agencies to provide appropriate contract oversight and management. In addition, a backlog of audits has caused DCAA to prioritize work on high dollar contracts, leaving unresolved many of the open audits of small businesses who are holding small dollar contracts.

This report provides several recommendations that the Panel believes will improve the defense business environment, reduce barriers to entry, spur innovation, increase competition, and aid in getting critical technology into the hands of the Warfighter.

PART I: THE DEFENSE INDUSTRIAL BASE

> "Creating the required future U.S. warfare capability, and thus protecting our national security, must be done by improving how the Department, including all its various component parts, does business; in order to support and sustain our position as the world's preeminent military power within current and expected resources levels."[2]
>
> David Walker
> Comptroller General of the United States, 1998-2008

The DOD relies on a broad industrial base to provide goods and services for everything from basic services (like facilities maintenance and security) to manufacturing to acquisitions of complex weapons systems. This broad industrial base, referred to as the Defense Industrial Base (DIB) is comprised of the private sector's worldwide industrial complex as well as U.S. Government and DOD capabilities. The DIB possesses the capabilities to perform sophisticated research and development (R&D), design, production, delivery, and maintenance of military weapons systems, subsystems, components, or parts to meet the needs of the military.[3] The DIB includes tens of thousands of prime contractors and subcontractors working under contract to DOD, and companies providing incidental materials and services to DOD, as well as government-owned/contractor-operated and government-owned/government-operated facilities. The DIB provides products and services that are essential to equipping and training the total force, as well as mobilizing, deploying, and sustaining military operations – but does not include the commercial infrastructure of providers of services such as power, communications, transportation, or utilities needed to meet military operational requirements.[4]

Defense Industrial Base Challenges

Sustaining Critical Sectors of the U.S. Defense Industrial Base

A recent study by the Center for Strategic and Budgetary Assessments (CSBA) entitled *Sustaining Critical Sectors of the U.S. Defense Industrial Base* found that "the United States has not pursued an overall strategy for preserving its domestic defense infrastructure since this industry emerged in the 1950s as a permanent sector of the U.S. economy."[5] The study concluded that "in light of the fiscal austerity likely to constrain U.S. spending on national security in the years ahead and the growing complexities and dangers in the nation's security environment, a long-term strategy seems imperative if the vital sectors of the defense industry are to be preserved."[6] One of the consequences of not having a strategy for maintaining the industrial base is that the United States could unwittingly lose the indigenous capability to develop and produce the critical goods and services that help maintain the superiority of the U.S. military. Many analysts believe that some key segments of the industrial base have already atrophied due to our lack of a long-term plan.

> "It is essential for the Pentagon and the U.S. Government writ large to develop a coherent long-term strategy for deciding what pieces of the defense industrial base are really going to be important going forward over the next several decades and make some real strategic choices."[7]
>
> Barry Watts
> Center for Strategic and Budgetary Assessments

Given the current fiscal environment and the pressure to reduce costs, it is more important than ever to take a strategic approach to maintaining the industrial base and ensure that short-term budget decisions do not undermine the long-term needs of the DIB. Discussing the importance of having a strategic plan for maintaining the industrial base, the CSBA report raised two main questions regarding the future of the U.S. Industrial Base:

1. Does the DIB function like a normal free market in which the forces of supply and demand dictate efficiency, innovation, and pricing?
2. What has been DOD's approach to sustain a strong DIB?

CSBA's Key Findings

Noted in their report, the CSBA found that the defense acquisition industry does not function like a free market system because of the government's dual role as both customer and regulator. The current defense acquisition system is particularly vulnerable to schedule delays and cost increases because cost control and timely delivery are often given a low priority. More often than not, changes to requirements and budgets also contribute to cost increases and schedule slips. CSBA further notes that reducing the number of units as a strategy to compensate for substantial cost growth has the effect of increasing the cost per unit (as well as potentially increasing the cost of other weapon systems built at the same facility) and generally gives the taxpayer far less value after investing large amounts of taxpayer money in R&D.

According to CSBA, the U.S. government has not made the effort necessary to develop a longterm strategy for assuring our DIB and the report contains a number of suggestions for improving the industrial base. For example, the report suggests that while using multiple prime contractors may not be possible because of consolidations that have already taken place in the defense industry, secondary contractors should be able to compete to improve efficiency at the component level. Other suggestions include:

- Providing firm initial operating capability dates for major defense acquisition projects and when necessary, descoping performance requirements to meet these dates.
- Examining certain segments of military businesses to see what capabilities are really necessary for today's and tomorrow's force; examine current and future mission possibilities to equip in a more strategic and efficient manner, e.g. an evolutionary strategy.

Improving the Strategic Relationship between the DOD and its Manufacturer and Service Suppliers

> "We need two things. We need budgets that produce programs that are profitable and that reach out to the talent we need, and we need an industrial base strategy that gives direction and predictability that the industry leaders need to make sound strategic business decisions."[8]
>
> Fred Downey
> Aerospace Industries Association

A July 2008 report by the Defense Business Board's (DBB) Task Group on a Strategic Relationship Model raised another concern regarding the industrial base, namely, that the level of dialogue between DOD's senior leadership and industry is inadequate. According to the report, communication between DOD and industry were more robust and effective in the past. For example, the Secretary of Defense and/or Deputy Secretary of Defense used to meet semiannually with key defense industry leaders to provide policy direction and guidance on key issues and to listen to industry concerns and suggestions. However, a number of factors have combined to contribute to the poor communication that now marks the relationship between DOD and the industrial base. These factors include

- DOD's lack of a clearly articulated view of the desired customer-supplier relationship;
- A narrow legal interpretation of allowable communications between customer and supplier; and
- A situation where most dialogue was limited to large manufacturing suppliers, while small and midsize companies in the industry were often ignored.

> "If we can give industry transparency through communications and [make that our desires] and our plans and back it up with execution clear to industry and to the financial community, we find that the markets tend to work and companies tend to get the investments they need."[9]
>
> Brett Lambert
> Deputy Assistant Secretary of Defense Manufacturing and Industrial Base Policy

Echoing some of the theme of the CSBA report, the DBB found that the lack of a clear strategic direction emanating from DOD has led to poorly aligned capital investment by industry. Without a healthy dialogue between industry, the combatant commanders, and the military services regarding future operational needs, it is difficult for suppliers to know where to invest their limited R&D resources. Without clear direction from DOD, industry's role in shaping military capability requirements is more likely to lead to technical "overreach" on requirements, and result in cost, schedule, and performance problems in acquisition programs. Restrictive laws and regulations contribute to this problem, making "management mobility" between industry and government difficult.[10]

> "If I can't talk to people and I can't get basic information, what happens? You are then forced to recruit retiring military people because that is the only way you get to understand what is going on inside, which then raises the specter of the issue of what is going on, and so I tighten laws about that, and then I get even more and more removed every step of the way until the point where I can't talk to my basic customer in order to understand what is going on."[11]
>
> Pierre Chao
> Senior Associate (Non-resident) International Security Program
> Center for Strategic and International Studies

Preparing the Industrial Base for the Twenty-First Century

Recently, Dr. Jacques Gansler published an extensive study of the challenges facing the DIB. Titled *Democracy's Arsenal: Creating a Twenty-First Century Defense Industry*, this study examined the full spectrum of challenges facing the DIB and suggests that in order to meet the defense challenges of the next century, the industrial structure must:

- Satisfy a broad range of mid-twenty-first century national security needs within the resources available;
- Be technologically advanced, and constantly evolve to maintain the United States' historic defense posture of technological superiority;
- Be highly innovative (in architectures, products, processes, and applications), focused on game changing technologies, and generate prototype demonstrations of these disruptive technologies;
- Remove barriers that now exist between civil and military integration, and to globalize the defense sector (while still recognizing that a few critical areas need to be protected) in order to benefit from the technological advances that are rapidly taking place in the commercial world and globally;
- Dramatically reduce the unit cost of weapon systems in order to be able to equip our forces with the quantity of equipment they need. The current cost of ships and planes are prohibitive. To procure the quantities required to execute future operations, unit costs must be decreased through improved product and process designs. Cost must be a firm military "requirement for all future weapons systems and systems of systems"
- Be highly competitive at all levels. There must be a least two firms in every critical area, but they need not all be domestically headquartered;
- Be agile and highly responsive. It must keep up with the adversary's changes and recognize that an adversary can acquire technology rapidly, on the global-technology market, and innovate in its use; and
- Be resilient enough to deal with the many forms of vulnerability that exist in today's environment (including physical and cyberattacks, natural disasters, fires, strikes, and changing geopolitical environments).

To achieve this standard, Dr. Gansler outlines nine actions that should be taken to create the 21st century DIB:

1. Focus on net-centric systems of systems - this shift from a platform-centric approach will have a major effect on planning, programming and budgeting processes, and will require new program management, systems engineering and architecture and governance structures which will take time for government and industry to adapt to.
2. Achieve lower costs, faster to-field times, and better performance.
3. Stay ahead by funding engines of innovation - in addition to increased investments for disruptive R&D, DOD should reestablish a separate allowable overhead expense for company-initiated independent research and development (IR&D). Congress should consider providing higher limits on the total size, the individual award amounts and the duration of DOD small business awards, particularly through the Small Business Innovation Research (SBIR)[12] program.
4. Conduct more best-value competitions (vs. the current trend to 'Low Price, Technically Acceptable)
5. Understand and realize the benefits of globalization.
6. Build a high-quality, high-skill government acquisition workforce.
7. Transform the DOD logistics system to a modern, world-class, information-based secure supply chain.
8. Recognize that contractors will represent a major portion of the total force in future large scale military operations (noting that in Iraq and Afghanistan it was over 50% of the total force)[13];
9. Specify and achieve a 21st century industrial structure - DOD, in conjunction with the Department of Homeland Security and the Intelligence Community, should articulate a clear vision and strategy for a 21st century national security industrial base.

   To achieve an integrated civil and military industrial organization, significant regulatory and legislative barriers need to be removed (such as government-unique cost accounting requirements; specialized military requirements; and unique government procurement regulations). Additionally, DOD profit and overhead policies need to encourage the structural shifts, capital investments, lower cost initiatives, and incentives for entry by new and commercial firms.

DOD's Approach to the Industrial Base

> "As the budget environment changes, we do expect some niche firms to face difficulty due to decreased demand. In such cases, we attempt to identify early warning signs through a variety of means to isolate and, if necessary, mitigate these issues, particularly if a firm offers truly critical, unique, and necessary capabilities."[14]
>
> Brett Lambert
> Deputy Assistant Secretary of Defense Manufacturing and Industrial Base Policy

The Office of Under Secretary of Defense for Acquisition, Technology, and Logistics released the *Annual Industrial Capabilities Report to Congress* in September 2011.[15]

As required by statute, the report includes a description of DOD's efforts to identify and address concerns regarding the nation's technology and industrial base, and to identify every initiative that is aimed at preserving or creating key segments of the technology and industrial base.[16] The report states

> During the past decade the Department relied on market forces to create, shape, and sustain the industrial, manufacturing, and technological capabilities in the industrial base intervening only when absolutely necessary to sustain essential defense capabilities. As the wars in Iraq and Afghanistan continue to evolve, and our nation continues to recover from the worst economic recession since the Great Depression, the Department faces significantly greater constraints on resources.
>
> These constraints will have significant impacts on the defense industrial base. The Department must work closely with our partners in the defense industry to ensure that we are better stewards of the taxpayers' money in these fiscally austere times. Our policies have changed to fit this environment.
>
> The fundamental starting point is the understanding that we in DOD do not make our weapons systems. They come from our defense industry. And these weapons systems are, second only to our superb men and women in uniform, what make our military power unrivaled and what provide the buttress of national and international security.
>
> A strong, technologically vibrant, and financially successful defense industry is therefore in the national interest. In this respect, the Warfighters' and taxpayers' interests are fundamentally aligned with those of the industry shareholders.

Next, the government's interest is not short-term, but long-term, like that of longterm investors. The Department will promote policies and actions that provide for long-term innovation, efficiency, profitability, and productivity growth.

The Department's report identifies seven guideposts in considering a "new era" for our defense industry. These guideposts state:

1. DOD will rely on normal market forces to make the most efficient adjustments to the DIB.
2. Competition will be one of the key drivers of productivity and value in all sectors of the economy, including defense. Therefore, the DOD will not be likely to support further consolidation of principal weapons systems prime contractors.
3. DOD will undertake a comprehensive study of the industry which will guide DOD as they seek to sustain the industrial base.
4. DOD's interest in the DIB extends throughout the spectrum and is focused on not only price prime contractors, but also on "lower tier" contractors.
5. DOD will give heightened attention to the importance of the "services" sector
6. DOD will seek to encourage new entrants.
7. DOD is committed to continue opening our markets while at the same time striking the appropriate balance with security concerns.

The Department has undertaken a new Sector-by-Sector, Tier-by-Tier (S2T2) effort to develop a repository of industrial base data.

According to Brett Lambert, Deputy Assistant Secretary of Defense Manufacturing and Industrial Base Policy, in testimony before the Panel, S2T2 will "also serve as a jumping off point for future assessments by all Defense Components, ensuring that data collection and analysis cumulates."[17]

Mr. Lambert went on to say that "sustaining and strengthening the data over time will also contribute to required insight to the Department's merger, acquisition, and divestiture reviews and other industrial base policies.

This information will also be used to manage our investments more effectively to ensure a healthy industrial base for those key sectors critical to future capabilities."[18]

Tools for Supporting the Defense Industrial Base

Defense Priorities and Allocations System and Special Priorities Assistance

Title I of the Defense Production Act (50 U.S.C. App. 2061 et seq.) provides the President the authority to require preferential performance on contracts and orders, as necessary, to meet national defense and emergency preparedness program requirements. Executive Order 12919 delegated these authorities to various federal departments and agencies and was revised by the President on March 16, 2011. The newly released executive order states that "the United States must have an industrial and technological base capable of meeting national defense requirements and capable of contributing to the technological superiority of its national defense equipment in peacetime and in times of national emergency. The domestic industrial and technological base is the foundation for national defense preparedness. The authorities provided in the Act shall be used to strengthen this base and to ensure it is capable of responding to the national defense needs of the United States.[19]

Previously, the Secretary of Commerce had authority to manage industrial resources. To implement its authority, the Department of Commerce (DOC) administered the Defense Priorities and Allocations System (DPAS). The DOC further delegated authority to the DOD under the DPAS to: (1) apply priority ratings to contracts and orders supporting national defense programs; and (2) request the DOC provide Special Priorities Assistance to resolve conflicts for industrial resources among both rated and unrated (i.e., non-defense) contracts and orders; and (3) authorize priority ratings for other U.S. federal agency and friendly-nation defense-related orders in the United States when such authorization furthers U.S. national interests. A DOD task force, known as the Priority Allocation of Industrial Resources (PAIR) task force, would convene to resolve industry constraints that interfere with military operations and Warfighter readiness. The task force works to ensure industrial resources are allocated to DOD programs in accordance with operational priorities when emergent requirements create competing demands among acquisition programs. It is too early to tell how the newly released executive order will be implemented and what, if any, changes will be made to the administration of DPAS and the conduct of the PAIR task force.

Title III of the Defense Production Act

The availability of domestic production capabilities for critical defense technologies is an essential element of national security. Title III of the

Defense Production Act (50 U.S.C. App. 2061 et seq.) is a program specifically designed to create, maintain, modernize, protect, expand, or restore industrial capabilities required for national defense. A key objective of the program created under Title III is to accelerate the transition of technologies from R&D to affordable production and insertion into defense systems. To create the needed industrial capacity, Title III authorities provide for the use of financial incentives in the form of purchases, purchase commitments, the purchase or lease of advanced manufacturing equipment for installation in government or privately owned facilities, the development of substitutes, and loans or loan guarantees. Title III activities strengthen the economic and technological competitiveness of the U.S. DIB and can reduce U.S. dependency on foreign sources of supply for critical materials and technologies. Title III is an authority, not a source of funds. Funding for individual Title III initiatives is provided by the joint or service program offices of record, defense agencies or other federal agencies as funding offsets for specific Title III efforts. Projects are developed in response to specific Government requirements and associated funding as provided for these efforts.

DOD Manufacturing Technology Program

For over 50 years, the DOD Manufacturing Technology (ManTech) Program has demonstrated its value through process technologies that make new products possible, as well as through manufacturing process improvements that get at the heart of defense system affordability challenges. The program provides the crucial links from technology invention to production of defense-critical needs that are in areas beyond the normal investment risk of industry. Ensuring that technology is affordable and producible remains key to making our forces more agile, deployable, sustainable, lethal, and dominant. While ManTech investments generally translate into initial system affordability improvements or cycle time reduction, investments are also made in new capabilities that provide dividends in system performance or life-cycle cost that can far outweigh the initial system delivery costs.

Industrial Base Innovation Fund

The Industrial Base Innovation Fund (IBIF) is a Congressionally-directed program established in 2008 and continues. The direction for IBIF was "to continue to make investments in manufacturing research that address DIB shortfalls especially related to surge production requirements and diminishing sources of defense material." In 2010 the IBIF Program made 21 contract awards totaling $20M. Small businesses accounted for 11 of those awards. IBIF made 16 awards to companies that are not weapon system integrators.

Recommendations

As discussed above, many recent studies and publications that have examined the DIB in order to gain insight into the complexity of the industrial base and the wide variety of factors influencing business behavior and the shaping of the DIB. Many of these studies highlight that the DIB is unique in that the DOD plays a role as both the customer and regulator of the DIB.

These studies also identify challenges to ensuring that the industrial base is positioned to support the needs of the nation in the 21st century. These challenges include the lack of a comprehensive DOD strategy for managing and maintaining an industrial base, the inconsistent communications that often marks the program office/private industry relationship, the fragmented nature of the industrial base. In light of these challenges, the Panel recommends:

RECOMMENDATION 1.1: Congress should require the Secretary of Defense, in coordination with the Service Secretaries, to develop a long-term strategy for maintaining a robust and effective DIB.

RECOMMENDATION 1.2: The Secretary of Defense should develop policies and mechanisms to improve communication with industry to better enable industry to allocate independent R&D resources and make investments to meet the future needs of the Department.

RECOMMENDATION 1.3: Congress should require the Secretary of Defense to provide a semiannual update to the congressional defense committees on the status of its S2T2 assessments and the utility of that data in shaping its policy toward the DIB.

PART II: THE USE OF INCENTIVES AND MANDATES TO SHAPE THE DEFENSE BUSINESS ENVIRONMENT

The Small Business Act of 1953 established that it is the declared policy of Congress that the Federal Government "should aid, counsel, assist, and protect, insofar as is possible, the interests of small business concerns in order to:

- preserve free competitive enterprise;
- ensure that a fair proportion of the total purchases and contracts or subcontracts for property and services for the Government (including but not limited to contracts or subcontracts for maintenance, repair, and construction) be placed with small-business enterprises;
- ensure that a fair proportion of the total sales of Government property be made to such enterprises; and
- maintain and strengthen the overall economy of the Nation."[20]

This policy is directly tied to the health of the DIB, for when the Small Business Act specifically addresses government contracting opportunities; it recognizes that it is "in the interest of maintaining or mobilizing the Nation's full productive capacity [and] . . . in the interest of war or national defense programs."[21] In addition to promoting free competitive enterprise, promoting fairness in federal procurement and the sales of government property, and strengthening the economy, economists and others have argued that the Federal Government should assist small businesses because small businesses are perceived to be at a disadvantage, compared with other businesses, in accessing capital and credit. In their view, lenders are less likely to lend to small businesses than to larger businesses because small businesses tend to be less established and have less credit history than larger businesses.[22] Also, lenders may be reluctant to lend to small businesses with innovative products because it might be difficult to collect enough reliable information to correctly estimate the risk for such products.[23]

Others have argued that assisting small businesses provides opportunities for minorities, women, and immigrants to increase their income and independence and to move into the economic mainstream of the American economy.[24] In their view, businesses owned by these demographic groups face even greater barriers in obtaining access to capital and credit than other small business owners due to discrimination and their higher likelihood of locating their business in a low or moderate income community. Operating a business in a low or moderate income community is often viewed by lenders as increasing the risk that the business owner will be unable to repay the loan.[25]

In recent years, interest in small businesses has increasingly focused on their role in job creation and retention. The Small Business Administration (SBA) has argued that small businesses play "a leading role as the driver of economic growth and job creation" in the national economy and note that "more than half of working Americans own or work for a small business and small businesses are responsible for two of every three net new private sector jobs created in recent years."[26] In addition, small businesses have led job formation during previous economic recoveries.[27] Thus the government has a real incentive to promote the integration of more small businesses into the defense industrial base.

Defining "Small Business"

By statute, a small business is one that is independently owned and not dominant in its field of operation, but the SBA is permitted to further define size standards on an industry by industry basis.[28]

In doing so, "the number of employees is used as one of the criteria in making such definition...the maximum number of employees which a small-business concern may have under the definition shall vary from industry to industry to the extent necessary to reflect differing characteristics of such industries and to take proper account of other relevant factors.[29]

Currently, there are over 1100 distinct industries recognized in the United States through the North American Industrial Classification System, plus additional sub industries. However, SBA is in the process of decreasing the number of size standards it will permit (from over 40 to 16) and combining industries together for the purposes of promulgating size standards. Further, until recently directed to do so by statute, SBA has not revisited the standards on a regular basis.[30] SBA has now formalized its process for establishing size standards, and in doing so considers: average firm size, start up costs and entry barriers, industry competition, distribution of firms by size, the effect of size standard revision on SBA's Federal contract assistance to small businesses.[31]

Unfortunately, by artificially limiting the number of size standards and combining industries for the purposes of setting standards, SBA's size standards do not accurately reflect the industries to which they are assigned. Most recently, when it published its professional, scientific and technical size standards, SBA excluded larger businesses from the definition of small, even though SBA's own analysis concluded that these firms were small.[32] In addition to harming small businesses, this harms the DOD's efforts to foster a strong industrial base, because small businesses may not be recognized as such, so DOD cannot provide them with the necessary opportunities or assistance.

Incentives and Mandates Related to Small Business

By statute, the SBA is required to use any SBA-derived industry specific size standards for determining whether a business is deemed small for federal procurement purposes. The SBA uses two measures to determine if a business is small for its business loan guaranty and investment programs: the aforementioned SBA-derived industry specific size standards or a combination of the business's net worth and net income. For example, the SBA's Small Business Investment Company (SBIC) program allows businesses to qualify as small if they meet the SBA's size standard for the industry in which the applicant is primarily engaged, or a separate financial size standard which has been established for the SBIC program. The SBIC's alternative size standard is currently set as a maximum net worth of no more than $18 million and average after-tax net income for the preceding two years of not more than $6 million. [33] Because the DOD's contracts typically involve a relatively large amount, many argue that there should be a dual size standard approach, as evidenced for the SBIC program, for determining whether a business is small for DOD contracting purposes is warranted. However, others argue that it is important for the Federal Government to maintain a consistent size standard regardless of the contracting agency.

While small business goals are negotiated with the SBA every two years with each department and agency, the government-wide prime contracting goal is currently 23 percent of all prime contract dollars. Given that federal spending through contracts is expected to decrease, there is a fear that the decrease will be disproportionately borne by small businesses. This decrease will affect the health of the industrial base, and will pose challenges to the economy as a whole.[34] To offset the effect of the decline in spending, the small business prime contracting goal should be raised.

Furthermore, allowing the DOD to have a goal of less than the government-wide goal means that small businesses may not have the same opportunities in some industries as they do in others, which in turn results in a less vibrant industrial base. For fiscal year 2010, SBA and DOD negotiated a goal of 22.8 percent,[35] however the agency accounted for 70 percent of federal prime contracts dollars.[36] Until 2009, SBA did not allow any agency to have a goal of less than the government-wide goal, because lower goals meant that each of the other agencies needed to contract at least 30 percent of their spend to small businesses to meet the goal, and would have denied small businesses opportunities in the very areas most crucial to the "war and defense programs" the Act is meant to address.[37]

Some may argue that raising the existing goal, which the DOD does not currently meet, will not result in the desired outcome. Much of what DOD purchases consist of investment in large systems such as ships, airplanes and vehicles that are beyond the scope that small businesses can provide. While more needs to be done to develop small business opportunities in the other areas of DOD business, such as construction contracts and information technology acquisitions, consideration could be given to adding some of these major defense acquisition programs to an exclusion list since these are not "practicable opportunities" for small businesses to serve as the prime contractor.

However, the Panel believes that there are many other recommendations in this report aimed at improving DOD's small business contracting performance that, if implemented properly, could enable DOD to successfully achieve a goal of greater than 23% small business participation on all prime contract dollars.

In the same way the DOD prime contract goal has been decreasing, the subcontracting goal has been decreasing as well. When the subcontracting goals were first established in 1978, they reflected Congressional concern that small businesses were receiving only 37.5 percent of subcontract dollars[38] – 1.6 percent more than the current goal. Indeed, the voluntary subcontracting goal established by the Executive Branch has been declining, from 40 percent in fiscal year 2003 to 35.9 in fiscal year 2010.

To ensure that small business subcontracting opportunities increase rather than decrease, the subcontracting goal should be returned to 40 percent.

A number of SBA and agency personnel currently are tasked, under various statutes and regulations, with protecting the interests of small businesses in the federal procurement process.

Among these personnel are procurement center representatives (PCRs), who are assigned by the SBA to work with the procuring activities in structuring acquisitions so as to maximize the participation of small businesses,[39] and Offices of Small and Disadvantaged Business Utilization (OSDBUs), which are established in the procuring activities to help devise alternatives to procurements involving "significant bundling," among other things.[40] However, the job descriptions for the OSDBU and PCR have not been updated since the major procurement reforms of the 1990s and 2000s and may not reflect the current acquisition and business environment.[41]

Current Challenges Facing Small Businesses

Small businesses were hit hard by the recent recession, accounting for almost 60% of the net job losses.[42] Although small firms added more than a half million jobs to the economy in fiscal year 2010, they continue to experience economic difficulties. For example, after examining recent economic data, the SBA found that

> While the number of startup establishments increased at the end of 2010, the number of employees at the average start-up continues to decline. This is not necessarily a business cycle issue; it may be a broader trend that contributed to the slow labor market expansion from the previous recession in 2001. Establishments that start smaller, on average, tend to stay smaller throughout their lives. Fortunately, the increase in the number of start-ups and expansions of existing businesses led to net employment increases in the last three quarters of 2010 (the latest period for which data are available). But in the fourth quarter of 2010, firms with fewer than 20 employees added only 38,000 net new jobs, while firms with more than 500 employees added 304,000. Even in the small business category of firms with 20-499 employees, new job creation decelerated from the previous quarter. It is not clear if this was a pullback from the strong third quarter of 2010, or the beginning of a new trend. The latest data on the number of start-ups is beginning to increase, and closures continue to decline. As one would imagine, the beginning of the downturn wiped out many shaky businesses. There was concern that such a long economic storm would overcome even stronger firms. It is encouraging that the latest data on establishment deaths and business bankruptcies show steady declines since 2009.[43]

The SBA concluded that "overall, the latest data for small business on job creation and start-ups indicate a continuing but choppy climb from the depths of the recession."[44] The following table provides net employment in the United States by firm size for fiscal year (FY) 2010.

In addition, the tightening of private sector lending standards for small businesses, which has recently eased somewhat but still remains "tight," has led to increased concern in Congress that small businesses might be prevented from accessing sufficient capital to start, continue, or expand their operations—actions which were expected to lead to higher levels of employment.[45] Others note that many federal agencies, including the DOD, failed to meet most of their small business federal contracting goals in fiscal year 2010.[46]

Table 1. Net Employment by Firm Size, Quarterly Data, Fiscal Year 2010

Firm Size	FY2010 Q1	FY2010 Q2	FY2010 Q3	FY2010 Q4	Total
<20 employees	-188,000	110,000	-34,000	38,000	-74,000
20-499 employees	-11,000	412,000	105,000	140,000	646,000
500+ employees	-71,000	166,000	50,000	304,000	449,000
Total	-311,000[a]	728,000[a]	151,000[a]	563,000[a]	1,131,000[a]

Source: U.S. Small Business Administration, Office of Advocacy, "Small Business Quarterly Bulletin: Second Quarter 2011".

Table 2. DOD Federal Contracting Goals and Percent of Fiscal Year 2010 Federal Contract Dollars Awarded to Small Businesses, by Type

Business Type	DOD Goal	DOD Percentage of FY2010 Federal Contract Dollars	Amount Awarded
Small Businesses	22.28%	20.94%	$61.12 billion
Small Disadvantaged Businesses	5.00%	7.12%	$20.77 billion
Women-Owned Small Businesses	5.00%	3.59%	$10.47 billion
HUBZone Small Businesses	3.00%	3.00%	$8.75 billion
Service-Disabled Veteran-Owned Small Businesses	3.00%	1.82%	$5.30 billion

Source: U.S. Small Business Administration, "Department of Defense: 2010 Small Business Procurement Scorecard."

Overall, in fiscal year 2010, the Federal Government awarded $537.8 billion in federal contracts. It met the 5% federal contracting goal for small disadvantaged businesses, but missed the goals for small businesses generally, women-owned small businesses, service-disabled veteran owned small businesses, and HUBZone small businesses.

Table 3. Federal Contracting Goals and Percentage of Fiscal Year 2010 Federal Contract Dollars Awarded to Small Businesses, by Type

Business Type	Federal Goal	Percentage of FY2010 Federal Contract Dollars	Amount Awarded
Small Businesses	23.0%	20.3%	$108.8 billion
Small Disadvantaged Businesses	5.0%	7.0%	$37.7 billion
Women-Owned Small Businesses	5.0%	4.4%	$23.5 billion
HUBZone Small Businesses	3.0%	2.4%	$12.7 billion
Service-Disabled Veteran-Owned Small Businesses	3.0%	2.2%	$12.1 billion

Source: U.S. Small Business Administration, "Government-Wide Performance: 2010 Small Business Procurement Scorecard," Washington, DC, http://www.sba.gov/sites/default/files/files/FY10%20SB%20Procurement%20 Scorecard_FINAL_GOVERNMENT%20WIDE.pdf (federal goals); and U.S. General Services Administration, Federal Procurement Data System—Next Generation, Washington, DC, https://www.fpds.gov/fpdsng/ (contract dollars).

Surety Bonds and Small Business Contracting

The SBA's Surety Bond Guarantee Program has been operational since April 1971.[47] It is designed to increase small business' access to federal, state, and local government contracting, as well as private sector contracting, by guaranteeing "bid, performance, and payment bonds for individual contracts of $2 million or less for small and emerging contractors who cannot obtain surety bonds through regular commercial channels."[48]

A surety bond is a three-party instrument between a surety (someone who agrees to be responsible for the debt or obligation of another), a contractor, and a project owner. The agreement binds the contractor to comply with the terms and conditions of a contract. If the contractor is unable to successfully perform the contract, the surety assumes the contractor's responsibilities and ensures that the project is completed. The surety bond reduces the risk of contracting.[49] As the performance portion of any surety bond serves to protect the taxpayer, and the payment portion of the bond protects subcontractors in need of payment, non-corporate sureties should be required to pledge specific and secure assets to be held by the government until performance and payment are complete.

Surety bonds are viewed as a means to encourage project owners to contract with small businesses which may not have the credit history or prior experience of larger businesses and are considered to be at greater risk of failing to comply with the contract's terms and conditions.[50] The three general types of surety bonds are:

- bid bonds guarantee that the bidder on a contract will enter into the contract and furnish the required payment and performance bonds if awarded the contract;
- payment bonds guarantee that suppliers and subcontractors will be paid for work performed under the contract; and
- performance bonds guarantee that the contractor will perform the contract in accordance with its terms and conditions.[51]

Surety bonds are important to small businesses interested in competing for a federal contract because the Federal Government requires prime contractors, prior to the award of a federal contract exceeding $150,000 for the construction, alteration or repair of any building or public work of the United States, to furnish a performance bond issued by a surety satisfactory to the officer awarding the contract, and in an amount the contracting officer considers adequate, to protect the Government. Prime contractors are also required to post a payment bond with a surety satisfactory to the contracting officer for the protection of all persons supplying labor and material in carrying out the work provided for in the contract. Both bonds become legally binding upon award of the contract and their "penal amounts," or the maximum amount of the surety's obligation, must generally be 100% of the original contract price plus 100% of any price increases.[52] Most state and local governments have adopted similar legislation, often called "Little Miller Acts," referencing the Miller Act of 1935 which established the federal requirement.[53] Many private project owners also require contractors to furnish a surety bond before awarding them a contract.

The program's sponsors argued in 1970 that "there is widespread evidence that a significant number of construction contracting organizations find varying degrees of difficulty in obtaining surety bonds`" and that "the major share of these organizations are small businesses, and many of them are headed by minority groups."[54] They argued that the Surety Bond Guarantee Program would "facilitate the entry and advancement of small and minority contractors in the construction business."[55] At that time, witnesses at congressional hearings testified that surety bonds were not necessarily required for most private sector construction contracts, but they were required for most public sector construction contracts.[56]

In fiscal year 2011, the SBA guaranteed 8,638 bid and final surety bonds (a payment bond, performance bond, or both a payment and performance bond) with a total contract value of about $3.6 billion.[57] Although the surety industry does not report the total value of the bonds it issues each year, estimates based on the total amount of premiums collected by the private sector in recent years suggests that the SBA's Surety Bond Guarantee Program represents, by design, a relatively small percentage of the market for surety bonds (from 0.6% to 2.5% of the value of surety bonds issued by the private sector).[58]

Excluding program costs of about $6 million annually, the SBA's surety bond program has had a positive cash flow in each of the last five fiscal years (see Table 4). For example, in fiscal year 2010, the program collected $9.2 million from fees and recoveries, paid out $4.3 million for claims, and had a net gain of $4.9 million.[59] In fiscal year 2011, the program collected $8.9 million from fees and recoveries, paid out $5.8 million for claims, and had a net gain of $3.1 million. There is currently about $42 million in the Surety Bond Guarantee Program Revolving Fund.[60]

Table 4. Surety Bond Guarantee Program, Net Cash Flow (excluding program costs of about $6 million annually)

Fiscal Year	Fees and Recoveries Collected	Claims Paid	Net Cash Flow
2007	$8.3 million	$5.2 million	$3.1 million
2008	$7.3 million	$6.6 million	$0.7 million
2009	$7.8 million	$6.0 million	$1.8 million
2010	$9.2 million	$4.3 million	$4.9 million
2011	$8.9 million	$5.8 million	$3.1 million

Source: U.S. Small Business Administration, Office of Congressional and Legislative Affairs.

Historically, the program's default rate has averaged about 2%. However, in 2011, the default rate increased to 3.7%.[61] According to the SBA, on average, the default rate on larger contracts tends to be lower than for smaller contracts and the recovery rate for larger contract defaults tends to be greater than for smaller contract defaults.[62]

There are currently 12 sureties participating in the Prior Approval Program and five sureties participating in the Preferred Surety Bond Guarantee

Program.[63] Agents empowered to represent a participating surety company are located, or licensed, in all 50 states and American Samoa, the District of Columbia, Guam, the Marshall Islands, the Northern Mariana Islands, Puerto Rico, and the Virgin Islands.[64]

Some organizations have argued that the program's $2 million bond limit should be increased to bring it more in line with the contracting amounts for other small business programs, such as the 8(a) Minority Small Business and Capital Ownership Development Program, the Historically Underutilized Business Zone (HUBZone) Program, the Women-Owned Small Business Federal Contract program, and the Service-Disabled Veteran-Owned Small Business Concerns Program. For example, under the HUBZone program federal contracting officials may provide a sole source award to a HUBZone certified small business if the anticipated award price of the contract will not exceed $6.5 million for manufacturing contracts or $4.0 million for other contract opportunities, and the contracting officer believes that the award can be made at a fair and reasonable price.[65]

The American Recovery and Reinvestment Act of 2009[66] temporarily increased, from February 17, 2009, through September 30, 2010, the maximum SBA surety bond amount from $2 million to $5 million, and allowed the amount to increase to $10 million if a federal contracting officer certified that the larger guarantee was "necessary." According to testimony by the SBA in front of the Committee on Small Business, this did not increase taxpayer costs.[67] Also, President Obama has recommended that the SBA surety bond program's maximum amount be increased to $5 million to "make it easier for small businesses to take advantage of contracting opportunities."[68] However, since construction contracts up to $6.5 million may be considered developmental under the Small Business Act, to the extent that the government's costs are covered by fees, SBA should be permitted to guarantee surety bonds up to that threshold.

Increasing the program's limit would increase the opportunities for small businesses to compete for federal contracts, especially in those departments, such as the Department of Defense, where the average size of construction contracts awarded to small businesses for fiscal year 2010 exceeded $5.9 million – nearly triple the size for which SBA can provide bonding support.[69]

Others have argued that the SBA's surety bond program should merge its Prior Approval Program and Preferred Surety Bond Guarantee Program into a single program featuring the more streamlined bond approval and monitoring processes under the Preferred Program. In the Prior Approval Program, the surety must obtain the SBA's approval before issuing a guaranteed bond.

Sureties selected to participate in the Preferred Surety Bond Guarantee Program may issue, monitor, and service SBA guaranteed bonds without SBA approval.[70]

Several industry groups, including the National Association of Surety Bond Producers and The Surety & Fidelity Association of America, have recommended that the programs be merged, the emphasis on reduced regulatory burdens under the Preferred Program be maintained, and the program's fees kept as low as economically feasible as a means to encourage more sureties to participate in the program.[71]

Contracting with Small Businesses Owned and Controlled by Minorities, Women, Service-Disabled Veterans and in Historically Underutilized Business Zones

In keeping with the "declared policy of the Congress that ... a fair proportion of the total purchases and contracts or subcontracts for property and services ... be placed with small-business enterprises,"[72] Congress has established government-wide goals for contracting with small businesses owned and controlled by minorities,[73] women, and service-disabled veterans and in HUBZones.[74]

It has also given agencies authority to conduct competitions in which only small business concerns which may include these types of businesses may compete (i.e., "set asides"), and to award such businesses contracts without competing them (i.e., sole-source awards).[75]

Agencies' performance in meeting their annual goals for the percentage of contract and subcontract dollars awarded to small businesses owned and controlled by minorities, women, service-disabled veterans, and in HUBZones is a perennial concern, and DOD, in particular, has often been criticized for failing to award more contracts to small businesses. As Table 5 illustrates, DOD's performance vis-à-vis its small business contracting goals has fluctuated from year to year, and it has repeatedly failed to meet its goals for contracting with women-owned, service-disabled veteran-owned small businesses, and businesses in HUBZones, in particular.

Moreover, some commentators have questioned whether the current goals provided for in statute—which were set over a decade ago—adequately reflect the availability of minority, women-owned, service-disabled veteran-owned small businesses, and HUBZone businesses in today's marketplace, and have advocated that these goals be increased.

Table 5. Percentage of Prime Contract and Dollars Awarded by DOD to Various Types of Small Businesses, Fiscal Year 2005-Fiscal Year 2010

Fiscal Year	Minority-owned	Women-owned	Service-disabled veteran-owned	HUBZone
FY2005	6.6% (3.4% 8(a))	3.0%	0.5%	1.9%
FY2006	6.3% (3.3% 8(a))	2.9%	0.7%	2.0%
FY2007	5.8% (2.4% 8(a))	2.9%	0.7%	2.2%
FY2008	6.1% (3.1% 8(a))	2.9%	1.0%	2.4%
FY2009	7.2% (4.4% 8(a))	3.4%	1.4%	3.3%
FY2010	7.1% (4.1% 8(a))	3.6%	1.8%	3.0%

Source: Small Business Goaling Reports for FY2005 through FY2010 as provided by the Department of Defense, *available at* https://www.fpds.gov/fpdsng_cms/index.php/reports.

Additional concerns have also been raised regarding the contracting programs for each of these types of small businesses. Specifically, with regards to the programs for minority-owned small businesses, there have been concerns about: (1) SBA's failure to recognize additional ethnic groups (e.g., Iranian Americans) as socially disadvantaged for purposes of the 8(a) Program;[76] (2) the differing rules governing participation in the 8(a) Program by small businesses owned by Alaska Native Corporations (ANCs) and by other small businesses; [77] (3) changes proposed to the FAR on September 9, 2011, in response to the decision by the U.S. Court of Appeals for the Federal Circuit in *Rothe Development Corporation v. Department of Defense*;[78] and (4) pending litigation challenging the constitutionality of the 8(a) Program.[79] Although authorized in 2000,[80] the set-asides for women-owned small businesses were not implemented until summer 2011, and it remains to be seen whether women-owned businesses raise concerns about the industries in which women-owned small businesses have been found to be "underrepresented" or "substantially underrepresented," or about agencies' inability to make sole-source awards to women-owned small businesses in circumstances when they could make sole-source awards to other types of small businesses. With service-disabled veteran-owned small businesses, there have been particular concerns about fraud because owners may self-certify their status as disabled veterans when participating in contracting programs under the Small Business Act (as opposed to those under the Veterans Benefits, Health Care, and Information Technology Act).[81] In addition, particularly where DOD contracts are concerned, some have objected to the absence of contracting goals or a set-aside program for small businesses owned by veterans who do not have a service-related disability.

Since 1991, DOD's Mentor-Protégé Program has offered substantial assistance to small disadvantaged businesses. Eligible mentor firms partner with a small disadvantaged business, a qualifying organization employing the severely disabled, and women-owned small business, a service disabled veteran owned small business, or a historically-underutilized business zone to successfully compete for prime contract and subcontract awards. [82] Many of the industry roundtable panelists lauded the DOD Mentor-Protégé program's ability to forge relationships between small businesses developing technology solutions with larger companies providing resources needed to manufacture and field products (see appendices). Vigorous encouragement from the USD(AT&L) and other DOD officials have helped reduce the negative perceptions of contracting with small businesses. In oral testimony before the Panel on November 1, 2012, Mr. Andre Gudger, Director of DOD's Small Business Program Office, cited the Mentor-Protégé program as "essential" for creating opportunities for small businesses.

Small Business Innovation Research Programs

In 1982, the Small Business Innovation Development Act (P.L. 97-219) established Small Business Innovation Research (SBIR) programs within Federal Agencies with extramural R&D budgets above specified thresholds. Congressional intent was to increase participation of small innovative companies in federally funded R&D.[83] As Government agencies with R&D budgets of $100 million or more are required to set aside a portion of these funds to finance the SBIR activity, the DOD is a major participant in the SBIR program.[84] Through FY2009, over 112,500 awards have been made totaling more than $26.9 billion across the Federal Government.[85] The program was last reauthorized through September 20, 2017 in the fiscal year 2012 NDAA (P.L. 112-81)[86]

Preferred Supplier Programs

The Panel sought discussion on the issue of competing as a small business in a very risk-averse acquisitions environment. They note the hesitancy of leadership within the acquisition community to allow their developing workforce to deviate from a rigid checklist and place a risk on a new technology from a small business. As discussed previously, program managers and contracting officers often lack access to timely and accurate market research that would allow them to consider small businesses as qualified and viable suppliers. A preferred supplier program based on past performance and other documented factors could aid defense acquisition officials in mitigating some of this risk aversion, but one could also argue that a preferred supplier program stifles competition and stands as a barrier to entry for new companies entering the market.

> "There is no doubt that the DOD acquisition community is very risk-averse, and we have to find ways to meter that risk-aversion and reduce the bureaucracy and leverage this critical sector of our economy to meet our national security requirements."[87]
>
> Mr. Joel L. Johnson
> Former Vice President, International Aerospace Industries Association of America, Inc.

Subcontracting as a Small Business

Recognizing that some prime contracts are unsuitable for performance by small businesses, Congress has also sought to promote small business' involvement as subcontractors or suppliers on federal contracts. It has done so, in part, by requiring that requiring that federal contracts valued in excess of $650,000 ($1.5 million for construction contracts) that offer subcontracting possibilities incorporate "subcontracting plans." Among other things, these plans are to include:

- "[s]eparate percentage goals" for subcontracting with various types of small businesses;
- a statement of the total dollars planned to be subcontracted and the total dollars planned to be subcontracted to small businesses;
- a listing of the principal types of supplies and services to be subcontracted; and
- assurances that the contractor will include terms relating to the government's policy of promoting contracting with small businesses in all subcontracts that offer subcontracting opportunities, and require all subcontractors receiving subcontracts valued in excess of $650,000 ($1.5 million for construction) that are not themselves small to adopt their own subcontracting plans.[88]

When the government restricts competition to small businesses, it has vested interest in seeing that a substantial portion of the work is actually performed by small businesses. Unfortunately, the rules governing the limitation on subcontracting are written in such a way that businesses attempting to comply are confused, and contracting personnel are hard pressed to determine if the rules are being followed. For example, in the case of a contract for services, "50 percent of the cost of contract performance incurred for personnel" must be spent by the small business on employees.[89] In order to track compliance, the prime contractor and the contracting officer would also need to know the cost basis of any subcontractors. Instead, limitations should be applied to price rather than cost, and subcontracts to similarly situated businesses should be allowed to be counted towards the prime performance requirements.[90] This simple change would allow greater transparency, ease compliance burdens for small businesses, allow contracting personnel to use the Federal Subcontracting Reporting System to track compliance, and encourage small businesses to team together to pursue larger contracts.

In addition, being a subcontractor (as opposed to a prime contractor) on a federal contract can leave small businesses at a distinct disadvantage because they lack "privity of contract" or a direct contractual relationship, with the Federal Government and, thus, have fewer protections than prime contractors. For example, the protections of the Prompt Payment Act generally do not extend to subcontractors on federal contracts,[91] whose payment is instead governed by whatever terms they negotiate with the prime contractor. Similarly, small businesses historically have been vulnerable to "bait and switch," when prime contractors represented to federal agencies that they would work with small business subcontractors in order to obtain federal contracts, but then failed to use these subcontractors after the contract was awarded.[92]

Additionally, small businesses operating as subcontractors often face difficulty in collecting payment in a timely manner. DOD recently amended the Defense Federal Acquisition Regulation (DFARS) to accelerate payments to all small businesses. Typically, invoices are paid by the payment terms of the contract. This is true even if the invoice has been approved and is ready for payment. However, it has long been the DOD's policy to pay small disadvantaged businesses as quickly as possible. The DFARS was amended to remove the word "disadvantaged" thereby, extending the policy uniformly to allow all small businesses to benefit from DOD's accelerated payment policy.[93]

The 111th Congress sought to address these unique difficulties of small business subcontractors when it enacted the Small Business Jobs Act (P.L. 111-240). Among other things, this act requires that prime contracts incorporating subcontracting plans also include terms obligating the contractor to:

> make a good faith effort to acquire articles, equipment, supplies, services, or materials, or obtain the performance of construction work from the small business concerns used in preparing and submitting ... the bid or proposal, in the same amount and quality used in preparing and submitting the bid or proposal, and provide the contracting officer with a written explanation whenever it fails to do so. In addition, the Act requires that prime contractors with subcontracting plans notify the contracting officer in writing if they pay a subcontractor a reduced price, or if payment is more than 90 days past due on a contract for which the federal agency has paid the prime contractor. This Act also requires that contracting officers consider any "unjustified failure" by a prime contractor to make full or timely payments to a subcontractor in evaluating the contractor's performance, and note any "history" of unjustified failures to make full or timely payment in the Federal Awardee Performance and Integrity Information System.[94]

The SBA recently proposed regulations that would implement these provisions of the Small Business Jobs Act. Among other things, SBA would: (1) define "subcontract" so as to exclude electricity and utilities, among other things, from the "subcontracting base" that is used in measuring contractors' performance in subcontracting; (2) authorize contracting officers to establish subcontracting goals reported as a percentage of total contract dollars (as well as a percentage of total subcontracting dollars); and (3) specify when a contractor could be said to have "used" a small business in preparing a bid or proposal.[95] Under the proposed regulations, the latter circumstances are fairly narrowly defined, encompassing only situations where the offeror specifically referenced a particular small business in its bid or proposal or entered into a written agreement with the small business for purposes of performing the specific contract as a subcontractor, or where the small business drafted portions of the proposal, or submitted pricing or technical information that appears in the bid or proposal, with the intent or understanding that the small business will perform that related work if the offeror is awarded a contract.[96] The SBA has also requested comments upon whether prime contractors who fail to meet their obligations in paying subcontractors should be required to enter into "funds control agreements" with third parties.[97]

It is too early to assess the effectiveness of the Small Business Jobs Act in addressing the concerns of small businesses. However, some concerns seem likely to persist, in large part, because the SBA expressly declined to adopt several findings or recommendations that the Government Accountability Office (GAO) had made for improving small business subcontracting.[98] Specifically, while GAO found that certain contractors had overstated their small business subcontracting achievements by excluding certain contracts, such as those for electricity and utilities, from the base, and recommended that prime contractors report their performance in subcontracting to small businesses as a percentage of total contract dollars (not subcontract dollars), SBA would exclude electricity and utility costs, among other things, when determining the amount available for subcontracting, and would continue to authorize reporting of subcontracting performance as a percentage of subcontract dollars. In addition, as the preamble to the proposed regulation notes, "[d]efining when a prime used a subcontractor in preparing a bid or proposal is very difficult,"[99] and some subcontractors may feel that they were "used" in preparing a bid or proposal notwithstanding the fact that their situations are not addressed by the proposed regulations. Likewise, the rule does not address the fact that many companies fail to file the mandatory subcontracting reports.[100] The lack of these reports skews the government-wide and agency specific reporting on subcontracting achievements.

Midsize Business Concerns

Federal law generally does not recognize "midsize businesses" as such, instead categorizes businesses as "small" and other than small. As a result, there is no standard definition of a "midsize business". However, some commentators have noted that businesses which are too large to qualify as "small" under the SBA's size standards, but are significantly smaller than the major defense contractors, experience unique difficulties in competing for DOD contracts. For example, after examining DOD contracting data from 1999 through 2009, the bipartisan, nonprofit Center for Strategic and International Studies (CSIS) concluded that midsize firms (defined by CSIS as firms which are too large to be categorized as small in the Federal Procurement Data System (FPDS) and having less than $3 billion in total annual revenue) were being "squeezed" out of DOD contracts by both large and small contractors.[101] From 1999 to 2009, CSIS found that the share of DOD contracts awarded to small firms increased (from 17.0% to 17.4%), the share awarded to midsize firms decreased (from 36.0% to 28.9%), and the share awarded to large firms increased (from 47.0% to 53.7%).[102] CSIS noted that the "squeezing out" of midsize firms was most noticeable in DOD contract awards for products, and somewhat less noticeable in contract awards for services and for R&D.[103]

Given the difficulties experienced by some midsize firms in DOD contracting, some have proposed that agencies be authorized to set aside contracts for which only midsize firms may compete, although doing so would arguably require defining what constitutes a midsize firm and would add complexity in terms of goal establishment and monitoring, set-aside determinations, and administrative processes. Others advocate programs that would allow midsize businesses to receive contracting preferences similar to those provided for small businesses, at least for a limited time. However, caution would be needed so as not to have unintended consequences for small businesses.

Contract Bundling

> "Bundling puts small businesses in a dependent subcontracting role, well-hidden from government decision-makers. It keeps us at arm's length from the government program managers who set the requirements. It also means that the government contracting leaders who make all the acquisition strategy decisions do not see the small businesses who are performing the work under the prime."[104]
>
> Linda Hillmer
> Chair, Small Business Division
> National Defense Industrial Association

Congress has imposed ever more stringent limitations upon the "bundling" or "consolidation" of contract requirements by federal agencies. "Bundling" entails:

> consolidate[ing] 2 or more procurement requirements for goods or services previously provided or performed under separate smaller contracts into a solicitation of offers for a single contract that is likely to be unsuitable for award to a small business concern due to—(A) the diversity, size, or specialized nature of the elements of performance specified; (B) the aggregate dollar value of the anticipated award; (C) the geographical dispersion of the contract performance sites; or (D) any combination of the factors described in subparagraphs (A), (B), and (C),[105]

while "consolidation" consists of the "use of a solicitation to obtain offers for a single contract or a multiple award contract to satisfy two or more requirements ... that have previously been provided ... or performed ... under two or more separate contracts smaller in cost than the total cost of the contract for which the offers are solicited."[106] The limitations on bundling and consolidation are intended to ensure that small businesses are not effectively excluded from contracting with the Federal Government, or relegated to serving as subcontractors or suppliers on federal contracts, as a result of how the Government structures its procurements. However, agencies are not prohibited from bundling or consolidation of requirements per se; rather, they are prohibited from "unnecessary and unjustified" bundling or consolidation, among other things.[107] It should be noted that the benefits of "bundling" to gain efficiencies and reduce costs has been realized under several contracts.

DOD is required to report all bundled contracts valued in excess of $5 million, as well as all contracts that involve consolidation of requirements valued in excess of $5.5 million, in FPDS,[108] and Table 6 gives the total number and value of procurements that DOD reported as bundled or consolidated in fiscal year 2005-2010.[109] However, because of the ways in which "bundling" and "consolidation" are defined for purposes of federal law, some groupings of contract requirements that are of concern to small businesses may not be reflected in Table 6, generally because the requirements are performed overseas (in the case of bundling) or were not previously provided or performed under separate smaller contracts. This is particularly the case where DOD procurements are concerned because DOD makes extensive use of indefinite delivery/indefinite quantity (ID/IQ) contracts that result in the issuance of orders for work that could potentially have been performed under separate smaller contracts. Additionally, the current definitions of bundling and consolidation fail to capture many construction contracts, since, due to the unique nature of each construction contract; the work is rarely if ever considered to be of a type previously performed.[110] This exclusion means that approximately $23 billion in DOD construction contracts were not required to be evaluated for contract bundling in fiscal year 2010.

The parallel analyses related to contract bundling and contract consolidation promote a distinction without a difference and should be eliminated.[111] Instead of overcomplicating bundling analyses, clear, simple guidance and definitions are needed. This guidance should protect small businesses against unjustified bundling, but not unduly burden the Department with reviews of work that will not ever be suitable for small businesses. For example, subjecting major weapon system procurements to bundling analyses does not add value for the Department or small businesses whereas changing the definition to capture construction projects would.

Table 6. DOD Bundled and Consolidated Procurements, Fiscal Year 2005--Fiscal Year 2010

Fiscal Year	Bundled Actions		Consolidated Actions	
	Number	Value	Number	Value
FY2005	1	$10 million	2	$14 million
FY2006	5	$160 million	124	$46.4 billion
FY2007	25	$1.6 billion	1,680	$90.7 billion
FY2008	16	$6.2 billion	31	$11.1 billion
FY2009	17	$1.6 billion	31	$1.8 billion
FY2010	63	$12.8 billion	161	$8.3 billion

Source: Federal Procurement Data System, Bundled and Consolidated Contracts Report, *available at* https://www.fpds.gov/fpdsng_cms.

Recommendations

As discussed above, small and midsize businesses face particular challenges in contracting with the DOD. In recent years the DOD has been unable to meet its small business federal procurement goals. Also, because the DOD's contracts tend to be fairly large, the SBA's limit of $2 million per contract for surety bonds is often not sufficient to be of use to small businesses contracting with the DOD. In addition, businesses that are other than small are excluded from set aside competitions under the authority of the Small Business Act. As a result, some small businesses that qualify for a set aside under the authority of the Small Business Act cannot obtain a required surety bond and some midsize businesses that can secure a required surety bond do not qualify for a set aside under the authority of the Small Business Act. Furthermore, the Panel found that DOD lacks the ability to track small business participation at the lower subcontract tiers. The Panel also found that DOD lacks a culture that fosters small business participation where appropriate. Based upon these findings, the Panel recommends the following actions:

RECOMMENDATION 2.1: Congress should consider increasing the DOD's small business prime contract and subcontract procurement goals, and increase accountability in the achievement of the procurement goals. Additionally, job descriptions for the OSDBU and PCR should be amended to reflect the challenges faced by these small business advocates and new procurement methodologies.

RECOMMENDATION 2.2: Congress should amend the Small Business Act to end the practice of combined size standards, stop placing artificial limits on the number of size standards, and better define what factors should be considered when size standards are proposed.[112]

RECOMMENDATION 2.3: Congress should increase the SBA's surety bond program limit from $2 million to $6.5 million government-wide or just for the DOD and improve the quality of bonds authorized.

RECOMMENDATION 2.4: Congress should require an independent assessment of the DOD's federal procurement contracting performance. The assessment should examine DOD's contracting procedures with the goal of making recommendations for increasing small business contracting with the DOD. The assessment should also examine the industrial make up of companies receiving subcontracts pursuant to the DOD Comprehensive Subcontracting Plan Test Program, with the goal of understanding how division or corporate subcontracting plans affect the subcontracting base. Additionally, the Comptroller General should be required include Small Business numbers and participation in their annual Assessment of Selected Weapons Programs report.

RECOMMENDATION 2.5: In developing the preferred supplier program, the Secretary of Defense should consider creating a mechanism for rewarding small businesses that have exhibited a history of superior contract performance. One approach could be to allow small businesses with a superior past performance record to sign multi-year contracts with DOD. Such a program could incentivize small companies to compete for and perform on government contracts by holding out the opportunity for locking in long-term contracts based on performance. Such a program could also incentivize small businesses to commit more of their limited resources to government projects or government related internal R&D funds because of the long-term payoff of a multiyear contract.

RECOMMENDATION 2.6: The Secretary of Defense should develop business processes and establish a single Department-wide information technology solution to comprehensively track subcontractor work down to appropriate levels and to track past performance of individual subcontractors at such level. Such an effort should develop methodology to optimize tracking so as to add value to the small business subcontracting process and accountability. This effort should not only capitalize on the OSD defense industrial databases that currently exist, but should also provide the information needed to conduct market research and identify critical component and/or service DIB issues.[113]

RECOMMENDATION 2.7: The Secretary of Defense should increase the accountability in achievement of the DOD's business procurement goals. On February 10, 2012 the Deputy Secretary of Defense issued a Department-wide Directive Type Memorandum that requires fiscal year 2012 Senior Executive Performance Plans to have a Small Business Performance Element. The Secretary should make this performance element permanent.[114]

RECOMMENDATION 2.8: The Secretary of Defense should examine the potential benefit of using contract incentives in prime contracts to increase small business subcontracting.

RECOMMENDATION 2.9: Congress should ensure that work being reserved for small business performance is actually performed by small businesses. Therefore, the FAR and DFARS should be reviewed, and clarified if required, to eliminate ambiguity regarding responsibility for ensuring compliance.

RECOMMENDATION 2.10: The Secretary of Defense should clarify and improve contract bundling decision processes to identify work likely to be suitable for small businesses, capture construction contracting, and improve transparency.

PART III: DEPARTMENT OF DEFENSE ACQUISITION ENVIRONMENT

Since the early days of the republic, the United States government has relied on contractors to provide goods and services to the military. Until World War II, the regulations and rules governing government contracting in general, and defense contracting in specific, were minimal. For example, in December 1907, the War Department issued a two page procurement notice for what some observers have called one of the most important government contracts in U.S. history: a contract to build a flying machine that is heavier than air. By the February 1908 deadline, the War Department received 41 proposals. The contract, awarded to the Orville and Wilbur Wright, is noteworthy for its brevity, focusing on engineering requirements and contractor compliance.[115] In contrast, according to a Boeing official, the original signed contract for the KC-46 tanker that was awarded to Boeing on February 24, 2011 consisted of 1,233 pages when originally signed– 70 pages of the basic contract, with references to 27 attachments consisting of an additional 1,163 pages.[116]

Starting in the early twentieth century, the number of regulations governing defense acquisitions, and the length of individual contracts (including mandatory clauses), began to grow. This trend increased substantially after World War II.[117] The growth in defense acquisition regulations was so rapid and uncoordinated that an Office of Federal Procurement Policy[118] study conducted in the late 1970s found that DOD had 79 different offices issuing procurement regulations, and that these offices developed a procurement process that consisted of some 30,000 pages of regulations.[119] This growth was fueled by a number of factors, including the increased complexity and specificity of military requirements, reform efforts that added more rules to the process, and increased inclusion of public policy goals into the acquisition process.

Concerned that the defense acquisitions process was an overly "complex and unwieldy system," Congress enacted the Federal Acquisition Streamlining Act of 1994 (P.L. 103-155) to overhaul the process.[120] Despite this act and various other congressional and executive branch efforts, many analysts believe that there is room to further streamline and simplify acquisition regulations in order to reduce the burdens on contractors.[121] As one observer noted, "If someone were asked to devise a contracting system for the federal government, it is inconceivable that one reasonable person or a committee of reasonable people could come up with our current system."[122]

These analysts believe that the complexity of the defense acquisition process, the constantly changing rules of the game, the regulatory burden associated with government contracts, and the business risks associated with defense contracts discourage many companies, particularly small and midsize companies, from competing for defense contracts.[123]

Organization of the Department of Defense

The Office of Manufacturing and Industrial Base Policy

Several offices in the DOD play a key role in shaping the defense business environment. The first is the Office of Manufacturing and Industrial Base Policy (MIBP) which exists within the Office of the Under Secretary of Defense for Acquisition, Technology, and Logistics (USD(AT&L)). The office was created "to sustain an environment that ensures the manufacturing and industrial base on which [DOD] depends is reliable, cost-effective, and sufficient to meet DOD requirements."[124] The Office's efforts are divided along two main axes, industrial base policy and programs (including the broad areas of global supply and industrial priority programs, mergers and acquisition review, and defense industrial financial analysis and metrics) and manufacturing (including industrial base assessments of various industry sectors, program execution of Defense Production Act authorities and the ManTech program). In setting policy and assessing the industrial base, MIPB is guided by several "guideposts":

- Primary reliance on market forces to make the most efficient adjustments to the DIB;
- Dependence on competition as one of the key drivers of productivity and value;
- Conduct an ongoing assessment of each sector of the DIB to understand its unique dynamics;
- Interest in the entire spectrum of the DIB, through all of its tiers;
- Encouragement to new entrants to the DIB by lowering barriers to entry; and
- Opening the defense market to the opportunities of globalization.

Small Business Offices within the Department of Defense

In addition to the MIBP, within the Office of the Secretary of Defense (OSD) and in each military department is an Office of Small Business Programs (OSBP) that reports directly to the Service Secretary and Under Secretary. Reporting to the Department's OSBP in each military service are small business field offices located in each major military command.

The OSBP is responsible for coordinating small business efforts for each of the military departments and field agencies. The Director of OSBP reports to the Under Secretary of Defense for Acquisition, Technology, and Logistics.

Although organizationally speaking the OSBP in both the military departments and in OSD have access to senior DOD decision makers, this access is often limited to crafting of policy; not implementation of existing policy. This is evident in OSBP's lack of participation in top management decision forums such as the Deputy Secretary's Management Action Group (DMAG), and the Defense Acquisition Board (DAB).

The OSPB describes its vision as creating "[a]n enabling environment in which the Department recognizes the value of and engages small businesses as critical suppliers of required Warfighting capabilities,"[125] and it seeks to fulfill this vision by implementing a range of programs and initiatives (e.g., Mentor-Protégé Program); providing training, conferences and events for small businesses and agency small business specialists; and working with the various Small Business Offices within DOD.

> "Just having the secretaries of the Services understand the importance of the small-business industrial base and embrace that and push that down through their services I think would make a huge difference."[126]
>
> Linda Hillmer
> Chair, Small Business Division
> National Defense Industrial Association

Deputy Secretary's Management Action Group (DMAG)

Also shaping the defense business environment is the Deputy's Management Action Group (DMAG). Recently formed by the Deputy Secretary of Defense the DMAGs role is to "ensure that management actions are synchronized and fully coordinated across the defense enterprise, is the forum for determining management actions in topics chosen by the Deputy Secretary, and will be the final decision point for the Strategic Choices Group, the program budget review, Compensation Task Force, Defense Business Systems Management Committee, Senior Oversight Committee and Efficiencies Initiative".[127] One of the principal integrated civilian-military governance bodies of DOD, the DMAG meets at the discretion of the Deputy Secretary of Defense to provide advice and assistance to the deputy on matters pertaining to DOD enterprise management, business transformation, and operations; and strategic-level coordination and integration of planning, programming, budgeting, execution, and assessment activities within the department.[128]

Defense Acquisition Board

The Defense Acquisition Board (DAB) is the Department's senior-level forum for advising the USD(AT&L) on critical decisions concerning defense acquisition programs. The DAB is composed of the DOD's senior executives and is chaired by the USD(AT&L). Other executive members of the DAB include:

- Vice Chairman, Joint Chiefs of Staff
- Under Secretary of Defense (Comptroller)
- Under Secretary of Defense (Policy)
- Under Secretary of Defense (Personnel and Readiness)

- Chief Information Officer, DOD
- Director of Operational Test and Evaluation
- Director, Cost Assessment and Program Evaluation
- Director, Acquisition Resources and Analysis (also executive secretary of the DAB)
- Secretaries of the Military Departments

Defense Acquisition Board advisors include the Assistant Secretary of Defense (Acquisition); Assistant Secretary of Defense (Logistics & Material Readiness): Deputy Under Secretary of Defense (Installations and Environment); Deputy Assistant Secretary of Defense for Manufacturing and Industrial Base Policy; DOD Deputy General Counsel (Acquisition & Logistics); Director, Defense Research & Engineering; DOD Component Acquisition Executives; the relevant OIPT Leader(s); Director, National Geospatial-Intelligence Agency; Deputy Director, Cost Assessment; Director, Defense Procurement & Acquisition Policy; Director, Systems Engineering; Director, Developmental Test & Evaluation; Director, Industrial Policy; Director International Cooperation; Assistant Secretary of Defense (Legislative Affairs); Chair, Functional Capabilities Board(s); and Cognizant Program Executive Officer(s).[129]

In addition to the DAB, the Information Technology (IT) Acquisition Board (ITAB) Review is convened to advise the USD(AT&L) or his or her designee on critical IT acquisition decisions, excluding defense business systems. These reviews are intended facilitate the accomplishment of the Department's acquisition-related responsibilities for IT.[130]

Surveying the Marketplace

Currently, program managers perform market research on a tactical, case-by-case, acquisition-specific basis using sources sought synopses and industry requests for information which, in effect, pushes the burden to industry to come back and identify themselves as being in the marketplace. This is typically burdensome for both large and small businesses. They must be "on their toes" in navigating the Federal Business Opportunities website and responding in a timely manner often requires having dedicated staff for this purpose only due to the Federal government's volume of posted opportunities and the all too frequent short-turn deadlines. For small businesses, it is especially difficult to respond timely with required acquisition-specific full capabilities packages which will be evaluated by the program managers to assess viability for small business set-aside competition determinations. If two or more capable small businesses are not identified as being qualified for the specific acquisition, full and open competition is generally pursued.

There are no tools or means available for program managers to perform the acquisition-specific market research on a strategic, enterprise basis unless they have a support contractor with the knowledge and commercial software tools available to them. The tactical, acquisition-specific approach to market research often reveals flawed outcomes which become more apparent as the proposed acquisition strategies move up through higher levels of command review and approval processes which are more strategic and enterprise focused. These flawed market research outcomes often adversely impact competition and small business opportunities, yet rework is often not a viable alternative due to long lead times associated with rework, Warfighter schedule impacts, and so forth. While small business specialists are generally cognizant of all market research activities across the local enterprise by virtue of their required role in the acquisition planning, strategy development, review and approval processes, they are insufficiently manned to engage early enough or to handle the capacity of acquisition-specific workload at their locations, and therefore, not able to effectively mitigate these recurring impacts to competition and small business (inclusive of small business subcontracting).

> "When a commercial company wants to do business with a small business, they do a lot of research on that small business, and they make sure that they are not only qualified, they are viable, they are able to perform. And that is the type of information that is missing from the market research that is available to Program Managers and small-business specialists and contracting officers right now."[131]
>
> Linda Hillmer
> Chair, Small Business Division
> National Defense Industrial Association

The Acquisition Workforce

The defense acquisition workforce, the men and women who are primarily responsible for the development and procurement of weapon systems, equipment, goods, and services for the Warfighter, has experienced a decline over nearly two decades and largely since the end of the Cold War. It is estimated that the overall total of defense acquisition workforce consists of 151,608 personnel, including 135,981 civilians and 15,627 military personnel.[132] Still, these estimates on the size of the defense acquisition workforce do not include private contractors, rendering the estimates incomplete. Given the increasing proportion of the defense procurement budget which is spent on service contracts, and the decline of the size of the military and DOD civilian workforces, there has been a rise in the amount of funds spent on contract labor to replace some civilian employees, however the significant rise of contract dollars is largely attributed to the period following 9/11.

As a result of the cuts to the acquisition workforce that took place in the 1990s, there were simply not enough acquisition personnel in the Department of Defense. Insufficient numbers of acquisition personnel increase the risk of poor contract planning, management, and oversight. DOD has made significant progress in increasing the capacity (size) of the acquisition workforce. Some analysts believe that the acquisition workforce has to be expanded even further. DOD has acknowledged that some limited additional growth may occur.

In some instances the problem is not the number, but the lack of expertise of acquisition personnel. According to DOD, the department is now turning to improving the capability of the workforce that is now in place."

During the administration of President George W. Bush, DOD saw a significant increase in the size, breadth and scope of service contracting; however, the size of the acquisition workforce continued to decline. A number of congressionally-directed audits, investigations, and hearings concluded that cost overruns for overhead and administrative costs, combined with alleged and documented cases of waste, fraud, and abuse, increased the overall contract cost, schedule and performance of DOD contracts. Some policymakers had questioned DOD's ability and capacity to manage such contracts, and whether DOD had the right mix of acquisition workforce personnel trained and equipped to oversee the size, scope and complexity of service contracts.[133]

> "Contracting officers are often overworked and under-equipped. Collaboration between program and contract staff is poor. And, there is a lot of confusion on what Government can say to industry and when."[134]
>
> Dr. Allan V. Burman
> President, Jefferson Solutions

In 2009, the Obama Administration first signaled a shift in policy with the goal of increasing the size of the defense acquisition workforce. At the same time, then-Secretary of Defense Robert Gates announced a new strategy to hire approximately 20,000 personnel through fiscal year 2015 and included this goal as part of the Administration's fiscal year 2011 budget request to Congress. However, DOD's plan to grow the workforce has been affected by then-Secretary Gates' announcement in August 2010 to end DOD in-sourcing because he found (as the Congressional Budget Office, Congressional Research Service, GAO and others have found) that competitive sourcing results in better cost savings and higher performance. Slow growth of the workforce may be affected as well by the current fiscal constraints imposed today on DOD.[135]

Current Workforce Challenges

> "We don't have the experience base in our defense acquisition workforce today. And if there is only one thing that this panel does, you have got to reinforce the need and the efforts to rebuild the capability of that workforce, because we rely on their judgments in making those contracting decisions."[136]
>
> David J. Berteau
> Senior Vice President and Director of International Security Program,
> Center for Strategic and International Studies

DOD faces a number of potential challenges to building a more effective acquisition workforce, including:

- The capability and capacity of the defense acquisition workforce
- Although DOD has made significant progress in increasing the size of the acquisition workforce, some argue that the DOD acquisition workforce lacks sufficient numbers of professionally qualified and skilled acquisition workforce personnel, and that this lack of skilled personnel has largely contributed to difficulty in the oversight and management of contracts and contractors. With the increasing size and complexity of DOD service contracts and the decline in the number and dearth of specific competency areas of DOD acquisition personnel, some officials have asserted that there are not enough DOD contracting officials available and experienced enough to manage the complexities of the new acquisition programs, or oversee private

sector contractors. In their absence, DOD has demonstrated an increased reliance (some would say over-reliance) on the use of private contractors. In a recent report which evaluated DOD's progress in rebuilding the acquisition workforce, DOD stated that:

> "Our work has found that a lack of an adequate number of trained acquisition and contract oversight personnel contributed to unmet expectations and has placed DOD, at times, at risk of potentially paying more than necessary."[137]

Dr. Gansler further points out, in his recent book, the undervaluing of the acquisition workforce---particularly in senior military positions. He notes that "in 1990, the army had five general officers with contracting background; and by 2007, all five of those positions had been eliminated."[138] Of note, the Air Force, in this same period, cut in half the number of general officers and senior executive services with contracting experience. Gansler further describes that this undervaluing "discouraged young military officers from going into the contracting field, since there were no longer general officer positions to which they could aspire." The Army has since implemented changes to help grow a senior level contracting workforce.

- The "retirement bulge" and the "bathtub effect"
- A large portion of senior DOD officials with a substantial number of years of experience are eligible to retire and may do so within the next few years; some members of this group have opted for buyouts and employment in the private sector. These personnel reductions may in all likelihood leave the DOD acquisition workforce with significant shortages in breadth, depth and senior experienced leadership. The Defense Contract Management Agency (DCMA) went from four general officers in 1990 to zero in 2009.
- DCMA had experienced significant reductions in the size of its acquisition workforce, as discussed in GAO's testimony before a subcommittee of the House Committee on Government Reform, as described here.

> I will now briefly touch on our work-related DCMA. By the early 2000s, DCMA had experience significant erosion of expertise, such that it could not fulfill all of its oversight functions. Since 2008, however, DCMA has been rebuilding its workforce, making increasing use of the Defense Acquisition Workforce Development Fund to do so.

For example, in fiscal year 2011, DCMA hired a little over 1,200 new employees under this authority. DCMA has also taken steps to rebuild their skill sets. For example, by the late 1990s DCMA had lost the majority of its contract cost price analysts. And as a result, DCMA reported that DOD's acquisitions were subject to unacceptable levels of cost risks.

Over the past two years, DCMA has hired almost 280 new contract cost price analysts and cost monitors.

Now, one challenge facing DCMA is its large percentage of retirement-eligible employees, making the agency vulnerable to the loss of valuable technical expertise and organizational knowledge. In part, DCMA plans to mitigate this risk through aggressive recruiting and bringing back retired annuitants (ph) to help raise the skill levels of the newer employees.[139]

- Since 2008, using authorities given through the Defense Acquisition Workforce Development Fund (DAWDF), DOD has made great strides to hire acquisition professionals. [140] The fund was established by Congress to "ensure that the Department of Defense acquisition workforce has the capacity, in both personnel and skills, needed to properly perform its mission, provide appropriate oversight of contractor performance, and ensure that the Department receives the best value for the expenditure of public resources.[141] As a result, a "new-hire bulge" now exists within the acquisition community. These parallel bulges constitute a "bathtub effect" as mid-career personnel are not abundant enough to adequately replace the retirement bulge, nor provide for enough on-hands mentorship to the new-hire bulge.
- Civilian personnel ceilings which limit the number of full-time federal employees
- The Office of Management and Budget (OMB) authorizes a specific number of federal positions that each agency must not exceed in each fiscal year. Once the limit is achieved, contractors are hired to fill the gaps. DOD does have exceptions and additional funding in place specifically for acquisition workforce hires that were initially funded with the DAWDF. These exceptions help to mitigate the impact of budgetary constraints on the size of the civilian workforce.[142] It should also be noted that for inherently governmental positions, the contractors are hired to support not fill.
- Federal pay freeze

- Some observers believe that current federal salary limits impact DOD's ability to attract the expertise needed for highly skilled and technical positions.
- Inconsistent definition of the acquisition workforce
- There are multiple views as to which occupational groupings and specific individuals do make up the defense acquisition workforce, making it difficult to determine just which positions are critical and most affected by the increased workloads and declining numbers of personnel. In addition, these definitions often do not include personnel who perform acquisition duties who are outside of the acquisition organizations. DOD is not required to identify these personnel nor is there a process to do so.[143]

Currently, the Department is improving the acquisition workforce certification process with greater emphasis on experience and being fully qualified.[144] Experience is a function of time and a key element for developing high quality employees. To ensure the acquisition workforce is fully qualified, all functional leaders have been asked to review their current functional experience and training requirements. For example, the certification experience requirement for the Systems Planning, Research, Development and Engineering - Program Systems Engineer career path has been expanded from 4 to 8 years.[145] This places greater emphasis on experience as a critical element in improving workforce quality and capability. In establishing experience requirements, emphasis will be placed on getting the right experiences to achieve both breadth and depth. Another example is the restructure of the Business career field into two distinct career paths, one for Cost Estimating and one for Financial Management. Most current acquisition career fields have three progressively higher levels of certification (I, II, or III); each level of certification possesses specific training, education, and experience requirements. For cost estimating, level III certified personnel must have completed level II certification and the following Defense Acquisition University (DAU) courses: Advanced Concepts in Cost Analysis; Software Cost Estimating; and Rates. Level III certification also requires:

- a Baccalaureate degree (any field of study)
- 3 semester credit hours from a calculus course
- 21 semester credit hours in any combination of the following fields of study: operations research, economics, mathematics, chemistry,

physics or other sciences where the utilization of advanced mathematical

- Skills in geometry, trigonometry, statistics, probability, and/or quantitative analysis
- 7 years of acquisition experience in Cost Estimating

Cost Estimating now requires 7 years of experience to achieve Level III, while Financial Management now requires 6 years.[146]

GAO has reported that DOD needed to assess the skills and competencies of and training provided to those people who have a role in acquisition but who are outside what DOD has formally defined as the acquisition workforce. GAO recommended that DOD establish criteria for identifying these personnel, assess the critical skills needed to perform their role in the acquisition process, and designate an organization that has the responsibility to track DOD's progress in identifying, developing, and overseeing personnel outside the defined acquisition workforce.[147]

In testimony before the HASC, DOD's top Human Capital Manager described efforts to improve workforce certification requirements, as described here.

- The Department is improving the acquisition workforce certification process with greater emphasis on experience and qualifications. For example, the engineering certification experience requirement was expanded from four to eight years. Leadership established cost estimating as a separate career path with increased education, training, and experience requirements and eight years of experience are now required to achieve Level III certification. These updated requirements reflect strong leadership emphasis on improved quality. Other quality initiatives under consideration include focus on how to improve development of talent after certification. While major investments in training are focused on early career certification, additional effort is needed to continue strengthening the mid-career workforce. Our goal is to have a workforce that is both fully certified to today's standards and also fully qualified to perform their duties as acquisition professionals. Our "certification-to-qualification" initiative will provide a critical fourth dimension to certification – on-the-job demonstration of mastery of functional competencies. This type of additional development effort for the mid-career workforce contributes to the objectives of the "recertification" requirement established by the Fiscal Year 2011 National Defense Authorization Act.[148]

These new career paths reflect strong leadership emphasis on increased training, education and experience elements for meeting certification standards. However, even with renewed emphasis on training the acquisition workforce, questions remain on whether or not the Department's acquisition professionals are equipped with the proper educational background or experience levels when compared with acquisition professionals in the defense industry. Based on the President's budget request for fiscal year, 2013, the Department is requesting $374 million in DADWF for recruiting and hiring acquisition and only $120 million, or less than a third, for training and development of the workforce.[149] This is in line with prior years' requests. However, the Department does increase spending on workforce recognition, retention, and assessment from $149 million in 2012 to a requested $450 million in 2013.[150] The budget overview document provided to Congress on February 13, 2012 states that that budget "supports continued strengthening of the acquisition workforce to ensure we achieve and sustain sufficient workforce capacity and capability" and notes that "training capacity has improved by approximately 19, 000 resident and 100, 000 online training seats per year."

Greater emphasis on training and experience are important factors in crafting a good acquisition professional. Equally important to the maturation of an acquisition professional is the education level achieved. Higher education is necessary because it equips acquisition professionals with complex skillsets in finance, systems engineering, logistics, and operations management needed to administer large contracts and manage long-term technology projects.

Recent Legislation Related to the Workforce

In order to address many of the changes facing the Department's workforce challenges, the Committee included eight provisions (Sections 931-938) in the National Defense Authorization Act (NDAA) for Fiscal Year 2012 (Public Law 112-81). While aimed at improving the Department's broader management of the total force (military, civilian, and contractor) the provisions should also serve to improve DOD's management of the acquisition workforce if implemented properly.

Recommendations

Some observers have argued that the defense acquisition workforce has struggled to manage and execute programs in the midst of challenges in acquiring and retaining a professionally certified and competent defense acquisition workforce. Many described the defense acquisition workforce as part of a "risk-averse" culture where program managers, in particular, have difficulty in balancing risks in managing cost, schedule and performance of acquisition programs. The DBB stated that:

> In addition, budget cuts and the increasing number of retirements have reduced the number, quality, and effectiveness of the Department of Defense civilian acquisition workforce and led to a greater reliance on contractors, which has reduced the Department's ability to lead and manage the acquisition life-cycle and created a risk-averse culture that undermines effective communication.[151]

DOD has, in some cases, outsourced program management and divested itself of critical skills that are difficult to develop – contracting officials, cost estimators, and systems engineers. This reliance on private contractors can create a potential conflict of interest and blur the lines between what work must be performed by federal employees and what work is permitted to be performed by private contractors. The Panel notes that just as it takes many years to develop a military leader capable of commanding at the senior ranks of the operational force, it takes a similar amount of time to develop an acquisition professional with the knowledge, skills, and experience needed to manage large defense acquisition efforts. In addition, the Panel found that constantly changing regulations lead to unnecessary complexity, confusion, and poor execution, only furthering challenges for the acquisition workforce.

The Panel also found that the DOD acquisition system lacks sufficient emphasis on small business participation. As a result of these concerns, the Panel recommends:

RECOMMENDATION 3.1: Continue to rebuild the breadth and depth of the defense acquisition workforce which will, in all likelihood, save money and improve acquisition outcomes.

RECOMMENDATION 3.2: Congress should closely monitor the Department's implementation of the Total Force Management provisions included in the NDAA for fiscal year 2012 with specific focus on the acquisition workforce.

RECOMMENDATION 3.3: Congress should require the Secretary of Defense to conduct an assessment of current certification and training requirements for the acquisition workforce and to seek legislative or regulatory changes targeted at improving the capability of the acquisition workforce.

RECOMMENDATION 3.4: Congress should require that the Director of the Office of Small Business Programs participate in senior-level decision forums such as the Defense Acquisition Board and Information Technology Advisory Board, and should regularly be invited to participate in the DMAG as matters relating to acquisitions and industrial base policy will arise.

RECOMMENDATION 3.5: The Secretary of Defense should conduct an assessment of the acquisition workforce to identify small business specialists, and take steps to address any deficiencies in capacity and capability of small business specialists. The assessment should examine how the small business specialists grew in relation to the acquisition workforce over the past five years. The Secretary should also consider establishing a small business specialist career field in the acquisition workforce and should also ensure all Department small business specialists report directly to the commander or deputy commander at all levels of the organization.

RECOMMENDATION 3.6: The Secretary of Defense should establish a requirement for formal coordination with the appropriate Office of Small Business Programs when certifying joint, as well as service-specific, requirements approval documents.

RECOMMENDATION 3.7: The Secretary of Defense should establish a requirement for formal coordination between senior acquisition officials and the appropriate Office of Small Business Programs when approving acquisition plans or strategies.

RECOMMENDATION 3.8: To improve its program management capacity as well as attract or develop personnel with critical skills in contract administration, cost estimation, and systems engineering, the Secretary of Defense should emphasize the professional development of its workforce by providing improvements to the Defense Acquisition University curriculum, emphasizing educational programs for post-graduate degrees in business or engineering, and mandating mid and senior level leaders complete appropriate contracting courses.

RECOMMENDATION 3.9: The Secretary of Defense should make small business programs an emphasis item when training the acquisition workforce. Currently, training on the Small Business Act contracting programs is provided by the Defense Acquisition University as a level 2 or level 3 elective course. Therefore, the training may be received only by contracting officers that likely to be managing the award of high dollar contracts that may, due to their scope or complexity, be unsuitable for small businesses. In contrast, lower level contracting personnel frequently handle those contracts most suitable for award to small businesses, but are not adequately trained on the programs.[152]

RECOMMENDATION 3.10: Congress should closely monitor the allocation of resources in general, and DAWDF specifically, to not only ensure an adequate number of acquisition personnel are available to the DOD, but to also maximize the ability to strengthen the capabilities (education, training and experience) of the acquisition workforce. Furthermore, the Secretary should consider creating a few high-quality, experienced teams of contracting specialists that can operate across programs to identify and mitigate contracting deficiencies.

RECOMMENDATION 3.12: The Secretary of Defense should systematically review the tasks small business specialists are required to perform and assess whether current activities are materially contributing to the achievement of agency small business goals and whether future activities should be more focused on providing small business market research, advice and reviews as early in the requirements approval process as feasible (no later than the requirement approval documentation process).

PART IV: BARRIERS TO TRANSITIONING TECHNOLOGY

> "The valley of death is a problem usually when we think about technology transition as a linear process of going from the research laboratory to some kind of demonstration until full-scale development or use in a commercial application. But it is not a linear process at all. Some people describe it as sausage making."[153]
>
> Dr. Stephen E. Cross
> Executive Vice President for Research Georgia Institute of Technology

Innovation is critical to the effective function of the national economy, as well as the defense establishment. As the National Research Council (NRC) noted, "a capacity to innovate and commercialize new high-technology products is increasingly a part of the international competition for economic leadership."[154] It is important to the note, though, that the process of developing innovative solutions is about more than just the development of new technologies. The NRC Committee on Comparative Innovation Policy defined innovation as:

> the transformation of an idea into a marketable product or services, a new or improved manufacturing or distribution process, or even a new method of providing a social service. The transformation involves an adaptive network of institutions that encompass a variety of informal and formal rules and procedures – a national innovation ecosystem – that shape how individuals and corporate entities create knowledge and collaborate to bring new products and services to market.[155]

This network of institutions, or national innovation ecosystem, is much broader than just industry. For the DOD, it includes industry as well as federal laboratories, universities, non-profit research institutions and federally funded research and development centers (FFRDCs). How they interact is based in part of how DOD funds its research activities, as well as special status given to some of these entities in the FAR.

The Role of Science and Technology in the DOD Innovation Ecosystem

Science and technology continue to play a major role in the military strategy of the United States.[156] Deploying a wide range of technologies gives the U.S. military a wide range of capabilities. The sophistication of that technology, along with the skills and doctrine needed to use it effectively and efficiently; helps provide the superiority and dominance on the battlefield which the U.S. military seeks to maintain against all adversaries. The role of science in this strategic effort is to provide the knowledge and understanding of the physical, chemical, biological, and human behavioral world needed to keep the technological frontiers moving forward.

The Department fiscally supports, directly and indirectly, much of the science and technology development upon which it depends. The majority of the direct support is allocated through the Research, Development, Test and Evaluation (RDT&E) program. For fiscal year 2012, Congress appropriated $72 billion for DOD's RDT&E program. DOD also indirectly supports the science and technology development it needs through the reimbursement of IR&D conducted by defense contractors with their own funds. Reimbursement occurs as an allowable cost on certain types of procurement contracts.[157]

The table below shows the funding for RDT&E over the last 42 fiscal years. Funding peaked in constant fiscal year 2011 dollars in fiscal year 2008 at $82 billion. While declining, the fiscal year 2012 figure remains higher than any time before fiscal year 2004.

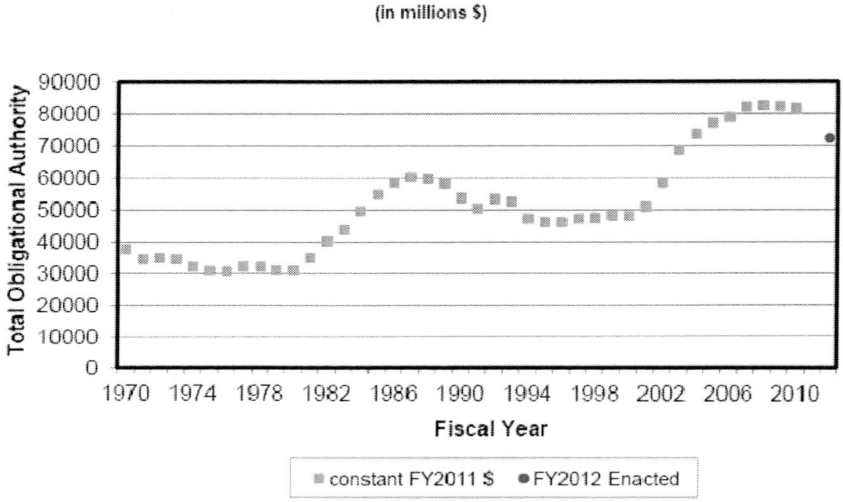

Source: Office of the Under Secretary of Defense (Comptroller), RDT&E Programs Budget Document, Office of the Secretary of Defense, various years (Congressional Research Service chart).

Figure 1. Department of Defense RDT&E Funding.

Most of the RDT&E funding is spent on the development of specific new military systems or the improvement of existing systems; for example, the development of the new Joint Strike Fighter aircraft or improvements to existing designs of F-16 or F-18 aircraft. Based on the fiscal year 2012 appropriation, 77% of RDT&E was devoted to this type of development. The development of specific military systems normally involves many man-years of iterative product and manufacturing design and development, tests, and evaluations. Normally about 15% of the RDT&E funding (17% based on the fiscal year 2012 appropriation), is spent on basic and applied research and the maturation and demonstration of new technologies that, in time, may lead to future military system or subsystem designs and manufacturing. Support for basic and applied research and technology maturation is referred to as DOD's Science and Technology (S&T) program. S&T projects, ranging from theoretical studies and bench-scale experiments to full scale (but not necessarily operational or production level) prototype systems, tend to be of shorter duration and involve less man-years of effort than the development of specific new military systems meant to be procured.

> "The valley of death is a problem usually when we think about technology transition as a linear process of going from the research laboratory to some kind of demonstration until full-scale development or use in a commercial application. But it is not a linear process at all. Some people describe it as sausage making."[158]
>
> Dr. Stephen E. Cross
> Executive Vice President for Research Georgia Institute of Technology

Basic research is especially important in this process of innovation, as it often leads to new areas of knowledge, such as new materials, sensors, nanotechnology, data extraction, etc, that in turn lead to new areas for development and commercial opportunity. As one book on the economics of the weapons acquisition process pointed out, "inadequate support of basic science not only may limit the future pace of weapons development, but may also explain many of the difficulties witnessed in recent weapons programs."[159] Since many of the industrial laboratories of the 1950's and 1960's have closed, or refocused to work on development activities, the predominance of basic research for DOD is carried out by the universities. That has in turn led to a trend of increased activities related to commercialization of technologies on university campuses to more quickly translate research into industrial products. For fiscal year 2012 Congress appropriated $12 billion for DOD's S&T program. The figure below shows the funding for the S&T program over time, in constant fiscal year 2011 dollars.

For the most part, the development of specific military systems is performed at private firms. A broader range of organizations, including private firms, both large and small, universities and other academic institutions, non-profit organizations, FFRDCs or University Affiliated Research Centers (UARCs)), and in-house DOD laboratories and product development centers, perform research and technology maturation.

Over half of the DOD's investment in basic research goes to universities. DOD also supports research and technology maturation in 67 DOD laboratories, 10 DOD-sponsored FFRDCs, and 13 DOD-sponsored UARCs.[160]

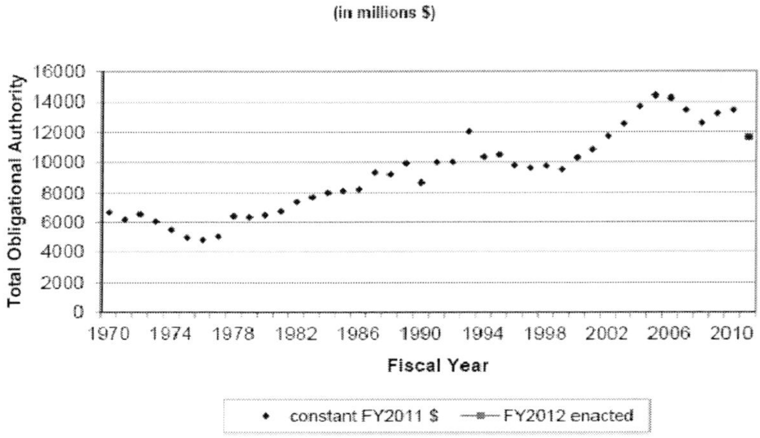

Source: Office of the Under Secretary of Defense (Comptroller), RDT&E Programs Budget Document, Office of the Secretary of Defense, various years (Congressional Research Service chart).

Figure 2. Funding for the S&T program within RDT&E Expenditures.

Support for research and technology maturation at private firms, universities, and non-profits is normally done through a competitive processes. However, FFRDCs and UARCs receive their funds non-competitively, although continued funding depends on DOD still needing their specialized services and they must compete on a regular schedule to retain the designation of FFRDC or UARC. DOD may also enter into cooperative R&D projects with other federal laboratories, universities, non-profits, and private firms. In a cooperative R&D project, a DOD laboratory will coordinate research with another R&D entity and share the results of their independently conducted and funded research.

Potential Impact of Future Funding Trends

"There are three big challenges facing the industry today. The first is the impact of the planned reductions and the budget reductions that are under way. The second is the importance to recognize that industry today, unlike industry in the past, has to remain competitive in the global financial markets. We can no longer rely on just the Federal Government to provide the funding for these companies. They have got to be competitive financially. And the third is where innovation is coming from in the 21st century, because we have a history of relying on defense contractors to come up with innovation."[161]

Mr. David J. Berteau
Senior Vice President and
Director of International Security Program Center for Strategic and International Studies

Faced with the prospect of hundreds of billions of dollars in additional cuts to its current 10-year budget as a result of the Budget Control Act of 2011 (P.L. 112-25), the DOD will be forced to balance its support for the RDT&E program budget and other parts of the budget. The recently announced Defense Strategic Guidance stated as overall defense spending is reduced, the department will protect and, in some cases, increase its investment in, among other things, new technologies associated with intelligence, surveillance, reconnaissance, unmanned systems, space, and cyberspace.[162] To the extent that fewer new large military platforms are called for in this new direction, one might expect the size of the system development part of the RDT&E budget to also decline. However, the need to continue developing new technologies could help maintain the S&T part of the budget. In the previous reduction of the defense budget after the end of the Cold War, Congress sought to sustain S&T funding, setting a goal of increasing the annual S&T funding 2% above inflation each year. If defense spending declines as expected, Congress may want to consider a similar commitment to sustaining S&T funding. Alternatively, Congress may wish to let S&T funding fall in proportion with the rest of the ODD budget, depending on the relative demands from, and priority given to, other parts of the budgets. Each option carries with it opportunity costs that must be heavily weighed. During this last budget reduction, the Congress and the department also sought to leverage its investment in science and technology development by engaging in more cooperative R&D with the commercial sector in those areas promising mutual benefit. Part of the logic of this policy was that DOD could benefit from higher volume production of dual-use products by the commercial sector. A set of dual-use-oriented programs (referred to as the Technology Reinvestment Program) were established to facilitate this cooperation. Congress later determined that these funds would better benefit the department if directed toward more military-unique projects.

"Another impediment to effective adoption of technology and innovation is the emphasis placed in acquisitions on 'lowest priced, technically acceptable.' The practice has become the preference and now forces behavior that overlooks better or more secure products to save on the price."[163]

Trey Hodgkins
Senior Vice President for National Security & Procurement Policy, TechAmerica

Government Laboratories and Technology Transfer

A recent report for the Office of Science and Technology Policy examined how the federal laboratories carry out technology transfer and commercialization activities.[164] While the study looked at more than just DOD laboratories, in the cross section examined, it found nine mutually influential factors that appear to affect the speed and extent of dissemination of technologies transferred from federal laboratories to the private sector.

- *Laboratory mission.* Technology transfer varies across laboratories due to the diversity and scope of their missions. Some laboratories are more inclined towards technology transfer that leads to commercialization because it is in the interest of achieving the mission of the laboratory, agency, or sub agency.
- *Laboratory management.* Differences between Government-Owned, Government-Operated (GOGO) and Government-Owned, Contractor-Operated (GOCO) laboratories can affect technology transfer and commercialization activities. GOCO laboratory leadership is often explicitly tasked to perform technology transfer and commercialization, while GOGO laboratories must comply with certain government regulations that do not affect GOCOs.
- *Congressional support and oversight.* Despite congressional support for technology transfer at the federal laboratories, congressional action and oversight can have the unintended consequence of encouraging a risk-averse culture towards technology transfer. Furthermore, technology transfer activities can be undermined when congressional priorities shift, as technology transfer requires long-term support.
- *Agency leadership and laboratory director support.* Support from agency leadership and laboratory directors can have a marked effect on technology transfer and commercialization activities. For example,

laboratory directors who support technology transfer may provide resources, flexibility, and creative license to their ORTAs. Those ORTAs who are not supported by their laboratory leadership can be severely constrained.

- *Organization and coordination of technology transfer and commercialization activities*. The centralization/decentralization of technology transfer functions at the agency and laboratory levels affects the speed of implementation of technology transfer actions, the consistency of policies across laboratories within an agency, and the ability to share best practices. The location of Offices of Research and Technology Applications (ORTAs) within an agency and laboratory can affect the visibility of technology transfer.
- *Offices of Research and Technology Applications*. Operations that seem to affect technology transfer and commercialization include the responsibilities of the office; the science, technology, and business expertise of the staff; the processes of the office; and the legal authorities available to the laboratory and how ORTA staff interpreted them.
- *Researchers*. Laboratory researchers, whose participation in technology transfer and commercialization processes varies across laboratories, may lack the knowledge, ability, and incentives necessary to undertake the research, administration, and business development involved in successful technology transfer.
- *Government-industry interactions*. Federal laboratories are not visible and accessible to industry, and certain regulations make it difficult for federal laboratories and industry to interact. According to partnership intermediaries, groups designed to broker partnerships between the laboratories and industry, industry is largely unaware of opportunities to collaborate with the federal laboratories.
- *Resources*. Resources devoted to technology transfer and commercialization vary across laboratories and agencies. Further, the extent to which the agencies and laboratories leverage federal, state, and local programs that support technology-based economic development may also affect technology transfer and commercialization.

The study also identified some innovative strategies believed to increase the speed and extent of dissemination of technology transfer that leads to commercialization. Although it was beyond the scope of the study to evaluate the effectiveness of these strategies, study interviewees suggested they could be useful to the laboratories or agencies as they pursue technology transfer and commercialization.

- Collaborate with universities.
- Increase laboratory director involvement in technology transfer activities.
- Strengthen or complement the skill set of the ORTA staff.
- Enhance education and incentives for researchers to engage in technology transfer.
- Use standardized agreements to streamline industry interactions.
- Increase visibility and access to federal laboratories by increasing outreach and use of partnership intermediaries. Increase availability of resources through leveraging economic development and commercialization programs and partnership intermediaries.

The Role of Universities, Research Institutions and Federally Funded Research and Development Centers (FFRDCs)

Besides the pure research role that many of the entities play, they also serve a number of other functions in the innovation ecosystem that support both the government's role, as well as supporting industry. For example, they can support the defense labs in many of the areas identified above. Additionally, they can be a source for meeting the workforce needs of the government.

Research at universities is a means for producing technical personnel in science, technology, engineering and mathematics (STEM) fields, which could be available to the government, but could also be recruited by industry as well. Additionally, partnerships between the government and universities, such as the strategic partnership between the University of Akron and DOD's Corrosion Prevention Office can result in successful education programs, directed research, and workforce training as well as outreach and policy development.

There are also mechanisms such as the Intergovernmental Personnel Act which allows for the targeted recruitment of researchers with expertise in specific areas on a limited basis and for government researchers to have a stint in university research positions.

Key Legislation Regarding Technology Transfer

> "The non-profit community would also like to have the opportunity to participate in programs like the UARCs, the University Affiliated Research Centers, for which we are currently excluded. If non-profits were given the opportunity to be designated as UARCs, then we would be able to much better support the STEM capabilities. Many of our organizations already provide internships, for example, to graduate students."[165]
>
> Dr. Norman Winarsky
> Vice President, Stanford Research Institute (SRI)

Beginning in 1980 with the Stevenson-Wydler Technology Innovation Act (P.L. 96-480) (Technology Innovation, Title 15 U.S. Code, §§3701 et seq. (2010)),[166] Congress has periodically passed legislation with the goal of increasing the federal laboratories' beneficial impact on society through technology transfer. The Stevenson-Wydler Act stated that the federal government shall strive, where appropriate, to transfer technology to state and local governments as well as to the private sector.[167] To facilitate the implementation of this mandate, it required that each laboratory with 200 or more technical staff have a technology transfer office, referred to as an ORTA.[168] The Bayh-Dole Act of 1980 (P.L. 96-517)[169] allowed federal agencies and GOGO laboratories to issue exclusive licenses to government-held patents. Previously only nonexclusive or open licenses could be granted. Subsequent amendments gave GOCO laboratories the same authority and allowed private companies to obtain an exclusive license for the full life of the government patent (not just five of the seventeen years as it had been previously authorized) (FLC 2009).

The Federal Technology Transfer Act (FTTA) of 1986 (P.L. 99-502) strengthened federal laboratory technology transfer through a mandate that technology transfer be a responsibility of all science and engineering professionals consistent with their mission responsibilities and the establishment of a principle of royalty sharing for federal inventors at a minimum of 15 percent.[170] The FTTA created a new mechanism for GOGO laboratories, whereby they could enter into Cooperative Research and Development Agreements (CRADAs) with other federal agencies, state or local governments, industrial organizations, and nonprofit organizations including universities. GOGO laboratories were also allowed to make advance agreements with large and small companies for patent or license rights to inventions resulting from CRADAs. The statute formalized the charter of the Federal Laboratory Consortium for Technology Transfer and required that each agency devote a fraction of their laboratory budget to this organization.[171] GOCO federal laboratories were granted the opportunity to enter into CRADAs and other activities with universities and private industry by the National Competitiveness Technology Transfer Act of 1989 (P.L. 101-189), under similar terms as stated by FTTA.

More recently, Congress has passed legislation to guarantee that a CRADA partner will receive a nonexclusive license at minimum (National Technology Transfer and Advancement Act of 1995 (P.L. 104-113)), revised the reporting requirement of technology transfer for the federal agencies (Technology Transfer Commercialization Act of 2000 (P.L. 106-404)), and required that the DOE establish a technology transfer coordinator position (Energy Policy Act of 2005 (P.L. 109-58)). In the report to accompany the NDAA for fiscal year 2012, the House Armed Services Committee included language criticizing the Department's technology transition initiatives.

> The committee understands that rapid acquisition programs are increasingly used in the place of dedicated technology transition programs and that the Department did not request any funds for fiscal year 2012 for the Defense Acquisition Challenge program. The committee is concerned about the effectiveness of technology transition within the Department and the opportunity to insert innovative and cost-saving technologies into Department of Defense acquisition programs.
>
> The committee notes that technology transition is essential to fulfilling the mandate of section 202 of the Weapon Systems Acquisition Reform Act of 2009 (Public Law 111-23), which requires acquisition strategies to ensure competition throughout the lifecycle of major defense acquisition programs. The committee believes that program managers are risk averse and are not incentivized to pull new technologies into programs of record in order to foster competition and reduce program cost. Consequently, there is a need for mechanisms external to a program of record to identify promising new technologies and to reduce the risk of technology transition for major defense acquisition programs.
>
> However, both the committee and the Government Accountability Office have observed that the Department's approach to funding transition is flawed and that multiple, small funding sources for specific transition activities offer a piecemeal solution to a more systemic problem. Accordingly, section 253 of the Duncan Hunter National Defense Authorization Act for Fiscal Year 2009 (Public Law 110-417) required the Under Secretary of Defense for Acquisition, Technology and Logistics (USD (AT&L)) to assess the feasibility of consolidating

technology transition accounts into one account to be managed at the Department-level. Section 253 also required the USD (AT&L) to submit a report to Congress on the aforementioned assessment and include recommendations concerning the streamlining and improvement of technology transition activities throughout the Department. Unfortunately, the USD(AT&L) has failed to comply with this statutory requirement. Therefore, the committee directs the Secretary of Defense to provide the congressional defense committees a briefing on the findings of the report required by section 253 by no later than June 30, 2011, so the committees can understand the full ramifications of the repeal or modification of technology transition and insertion activities, such as the Technology Transition Initiative and the Defense Acquisition Challenge program.

Recommendations

The Panel found that DOD lacks a clearly articulated strategy that would provide a corporate vision of DOD's future technology needs. Starting in 1989, DOD was required to submit a Critical Technologies Plan that eventually morphed into the Defense Technology Area Plan and then supported by the Joint Warfighting Science and Technology Plan and the Defense Technology Objectives. Over the years, these requirements have been decreased resulting in a lack of strategy.

This lack of a strategy, fully coordinated throughout the various military departments, combatant commands and defense agencies, makes it difficult for industry to have visibility into the future developmental needs of the Department, and makes it challenging for both industry and the government to make important investment decisions, such as those needed for internal IR&D and management of the supply chain.

The Panel also found that DOD acquisition policies and processes provide little incentive (and in some cases, disincentives) for defense industry to invest in innovation.

Additionally, the Panel found that DOD has a multiplicity of uncoordinated funding mechanisms for technology transition to get early stage R&D efforts across the "valley of death."

The GAO identified more than 20 initiatives for rapidly satisfying urgent operational needs from the Warfighter. In addition, there are a number of technology transition mechanisms such as Defense Production Act Title III, ManTech, SBIR program, and the IBIF that have been implemented to serve slightly different but related purposes. Recent efforts to speed technology to the battlefields in Iraq and Afghanistan, such as the Rapid Innovation Fund (RIF), have also been added.

The GAO observed that the reach of these initiatives is limited and there is no unified, corporate approach to using them and noted that the Department's approach to funding transition is flawed and that multiple, small funding sources for specific transition activities offer a piecemeal solution to a more systemic problem.[172]

In response to those observations, the HASC requested a report from DOD in the fiscal year 2009 NDAA (and again in the fiscal year 2012 NDAA) on DOD plans to streamline these processes, but that plan has not yet been delivered. In light of these findings, the Panel makes the following recommendations:

RECOMMENDATION 4.1: Congress should direct the Secretary of Defense to develop and issue a technology strategy that would provide visibility into future investment areas aligned to DOD needs and capability gaps. Such a technology plan should be synchronized with the industrial base strategy recommended in Recommendation 1.1.

RECOMMENDATION 4.2: The Secretary of Defense should immediately comply with Section 234 of the NDAA for fiscal year 2012 and report back on plans for streamlining the rapid acquisition and technology transition processes of the DOD. As part of that review, the Secretary should assess the effectiveness of external support mechanisms, as well as how to leverage state and local economic development and commercialization programs.

RECOMMENDATION 4.3: The Secretary of Defense should institute a more robust process for technology scouting and market research that could give the government greater visibility into the global market. Such an effort could also provide a more central point of access into DOD to allow industry to understand DOD needs. The Panel is aware of many recent efforts such as the Open Business Cell, the Technology Support Working Group and the Defense Venture Catalyst Initiative that could be leveraged in developing that technology scouting process. This recommendation includes reviewing and collaborating, as appropriate, on SBIR projects at the Department of Energy, National Science Foundation, and other civilian agencies to consider military applications of other SBIR projects. The small business program offices in the Military Departments and at OSD should also be leveraged to assist in acquiring and distributing market research.

RECOMMENDATION 4.4: Congress should require an analysis of major acquisition programs to determine how successful they have been in leveraging technology developed by small businesses, such as those developed and commercialized through the SBIR program. This review should also develop ongoing measures of effectiveness and performance that could provide empirical data on how well the DOD is doing in integrating small business technologies into major acquisitions.

RECOMMENDATION 4.5: For appropriate weapon system development efforts, the Secretary of Defense should explore creating a mechanism for maintaining consistent levels of R&D funding that focus on technology developments directly related to next generation system or material solutions. Such a mechanism for sustained R&D funding could substantially improve the health and responsiveness of the DIB by reducing instability, and could help prevent technologies from being prematurely included in an acquisition effort, thereby reducing the problems resulting from concurrent system development and production.

RECOMMENDATION 4.6: The Secretary of Defense should facilitate increased partnering between non-profit or for-profit organizations with expertise in R&D and universities to assist the universities in advancing basic research efforts. The Panel believes that such partnerships could enable the Department to not only acquire a larger portfolio of technologies to enhance the country's national security, but ensures the Pentagon gets the best return on its investments in the R&D accounts.

RECOMMENDATION 4.7: The Secretary of Defense should evaluate the entire research and engineering infrastructure of the Department, including all of the laboratories, test and evaluation infrastructure, product and logistics centers, depots, and arsenals, to understand how these resources might be better utilized to support technology development, transfer and integration. For example, a consolidated effort between the DOD laboratories and the depots and arsenals could increase the manufacturing capacity to execute rapid prototyping. Such an effort could enable small businesses to leverage existing Government infrastructure demonstrate small scale capabilities and refine manufacturing processes before scaling up for commercial production.

RECOMMENDATION 4.8: Congress should examine the uncoordinated funding mechanisms for technology transition and take steps to streamline funding mechanisms to increase transparency, reduce redundancy of effort and target technology development (both basic and applied research) to meet Warfighter requirements.

PART V: NAVIGATING THE DEFENSE ACQUISITION SYSTEM

> "Smaller companies...they don't even begin to know where to start, how to interface, how to begin to get in. They would love to have an ombudsman or somebody that could be their champion."[173]
>
> Pierre Chao
> Senior Associate (Non-resident) International Security Program
> Center for Strategic and International Studies

DOD procurement activities are governed by three sets of federal government regulations. The first set of regulations, which apply to the entire federal government (including DOD unless stated otherwise), are found in the FAR. The second set of regulations, apply only to DOD and are found in the DOD supplement referred to as the DFARS. The third set of regulations, apply only to individual DOD components and are found in component-unique FAR supplements.[174]

Procurement actions in DOD must adhere to the various regulations, including those executed as part of DOD's acquisition programs, and both contracting officers and program managers must take the regulations into account during the planning and execution of contracts and programs. The goals of the current federal acquisition process are to deliver to the government "the best value product or service... while maintaining the public's trust and fulfilling public policy objectives."[175] Reflecting these goals, changes that have been made to the acquisition process over the years have primarily been aimed at improving the acquisition process or promoting public policy goals. Examples of regulations that reflect public policy goals include the requirement to purchase certain goods from domestic suppliers (such as the Berry Amendment and Buy America Act), requirements to take steps to combat trafficking in persons, set asides to promote small businesses and other entities perceived as disadvantaged, and the International Traffic in Arms Regulation (ITAR).

In some instances, the goals of obtaining the best value for the government and promoting public policy goals are in conflict with one another. For example, the public policy goal of requiring certain items to be manufactured domestically could increase the cost to the government.

Complexity of the Acquisition System

Contracting with the federal government is a highly regulated process governed by a myriad of statutes and regulations. These regulations govern such issues as how DOD solicits, negotiates, and awards a contract; what costs DOD will reimburse and how contractors must account for those costs; the information systems used by contractors; and how contractors must comply with rules regarding such socio-economic goals as affirmative action, trafficking in persons, and maintaining a drug-free workplace.[176] The complexity of the regulations can make it difficult for some companies to enter the government contracting arena. As one observer noted, "contracting with the federal government is a highly regulated process with many traps for the unsuspecting."[177]

> "Long procurement lead times typically encountered at DOD are also a barrier, particularly for commercial companies. They are unaccustomed to such long lead times and usually operate in environments using agile development in incremental models for short cycles of 6 months to a year. The DOD's 24-or-more-month lead times are not conducive to attracting the innovation these companies could bring to bear."[178]
>
> Trey Hodgkins
> Senior Vice President
> National Security & Procurement Policy
> TechAmerica

A number of analysts argue that the complexity of the acquisition system dissuades a number of companies from competing for government contracts.[179] Small and midsize businesses, which often do not have the resources to hire in-house counsel or experts in government contracting, may find government contracting too difficult to navigate. Not only is the defense acquisition process complex, defense acquisition rules are constantly changing, making it challenging for companies to keep up with changes that can impact their business.

Constantly Changing Rules of the Game

The rules and regulations governing defense acquisitions are constantly changing and evolving. Sometimes the rules change to keep pace with technology and manufacturing innovations. For example, the acquisition regulations at the turn of the twentieth century were created before the development of complex information technology systems. Regulations also change to reflect the public policy priorities of government. But the most common reason for changes to defense acquisition regulations is a virtually uninterrupted cycle of reform aimed at improving the effectiveness of the defense acquisition system.

For more than 100 years, the executive and legislative branches have been frustrated with the perceived level of mismanagement and corruption in defense acquisitions and have spent significant resources seeking to reform and improve the process. [180] Over the last fifty years, both Congress and DOD have been active in trying to improve defense acquisitions through legislation and regulatory changes. As then Deputy Secretary of Defense William Lynn stated, "Since the end of World War II, there have been nearly 130 studies on acquisition reform."[181] The result of the ongoing effort to reform defense acquisitions is a near-constant revision and amendment of the statutes and regulations that govern DOD procurements are constantly being revised and amended.

The ever-changing nature of the laws and regulations governing defense acquisitions can make it difficult for companies with limited resources to stay abreast of the changes that could impact their contracts of business strategies. The extent of legislative and regulatory change has fueled a cottage industry dedicated to helping businesses stay informed of the most recent changes. Every year, books, seminars, and webinars are aimed at keeping business owners abreast of changes that could affect their business. Some analysts believe that companies may shy away from government contracts out of concern that the contracting rules could be changed in the middle of the game, making it more risky to pursue a business strategy geared towards winning government contracts.

Legislative Changes

In recent years, the primary mechanism in which Congress has exercised its legislative powers to reform defense acquisitions has been the annual NDAA. Sections of these acts have prescribed requirements applicable to both specific acquisition programs and the acquisition structure overall, the latter of which has typically been addressed in Title VIII of the bill, which is usually called "Acquisition Policy, Acquisition Management, and Related Matters." Over the last five years, this Title in the NDAA dealing with acquisitions included more than 240 sections.[182] In addition to the NDAA, Congress has also chosen to enact legislation affecting defense acquisitions in a stand-alone bill. For example, in May 2009, Congress passed and the President signed into law the Weapon Systems Acquisition Reform Act of 2009 (P.L. 111- 23), which addressed a variety of issues such as competition and conflicts of interest in defense contracting.

Regulatory Changes

The rules and regulations governing defense acquisitions can also change at a rapid pace. For example, the DOD regulations on acquisition, referred to as DOD Directive 5000, or simply the 5000 series, was established in 1971. Over the next 40 years, the process for acquiring weapon systems set forth in the 5000 series was revised more than a dozen times – representing a change approximately once every three years. In some cases, the changes have been dramatic. The 5000 series have been issued and reissued, with changed page counts ranging from as few as eight to as many as 840. 2007 marks the most recent certification of these documents.[183]

Assessing the Acquisition System

> "As long as we have a requirements and acquisition system that takes 20, 25 years to get something from concept to actually out in the field, you are going to have always a fundamental disconnect between strategy and what we are buying."[184]
>
> Pierre Chao
> Senior Associate (Non-resident) International Security Program
> Center for Strategic and International Studies

In April 2009, the Defense Science Board issued a report titled, "Creating a DOD Strategic Acquisition Platform".[185] Among its findings, the report noted:

- Fixing the acquisition process is a national security issue, as the current process takes too long to produce weapons that are too expensive and often technically outdated by the time they are fielded.
- The acquisition of services receives far less attention than that of materiel, yet it is a growing part of the defense budget noting that in 2010 it was 57% of the DOD's acquisition budget.
- Many problems appear to be caused by the use of immature technology, requirements "creep," or funding instability. Such problems are really only symptoms of the lack of experienced judgment on the part of DOD acquisition personnel, and others involved in the process.

- Many organizations in DOD are not aligned with departmental acquisition goals and objectives.

This study went on to recommend four critical elements to achieving better acquisition management:

- ***Buy the right things.*** The DOD has a weak analytical foundation for making decisions on "what to buy" and focuses too heavily on "how to buy." The Secretary of Defense should reform the strategic military planning system and establish a genuine business plan for DOD to discipline resource allocation. This plan should comprehensively address the national security objectives of the DOD, as well as the human and financial resources needed to accomplish them.
- ***Select an effective leadership team.*** The Department must hire and assign individuals with proven track records of acquisition success, and should consider not starting programs if the proper experienced personnel are not available.
- ***Reform and streamline the acquisition process.*** The goal of reform should be to dramatically reduce the time between identifying a new operational need and fielding operationally useful equipment. The acquisition of commercial or commercially derived products presents a significant opportunity to DOD to achieve these goals, but requires a different mindset and management approach. Lack of experience in working with commercial products as well as the need to better assess tradeoffs in system requirements between military standard and commercial grade contribute to those difficulties.
- ***Improve acquisition execution.*** The Department should implement the following practices: change the concept of "requirements" to "capabilities;" manage technology development portfolios and create contingency plans for technology insertion; maintain persistent technology demonstration prototypes to use as technology demonstrators; ensure technology readiness before planned insertion; use competitive prototypes when possible; use spiral development and block upgrades with stable capabilities for each block; give program mangers capabilities and performance tradeoff authority. In addition to those practices, the DOD should ensure proper staffing and support a high-quality, coherent and competent workforce, as well as develop and implement a system of performance metrics that can be used to

monitor uniformly the progress of a program (and reward or penalize acquisition professionals according to those metrics).

Management, Oversight and Audit Agencies

> "Contract auditors measure their success by the numbers of costs that are questioned and the amount of those questioned costs that are sustained. What they really ought to measure their success by is the timeliness and value of the ultimate delivery of the results of those contracts."[186]
>
> Mr. David J. Berteau
> Senior Vice President and Director of International Security Program
> Center for Strategic and International Studies

In addition to the financial cost of complying with some regulations, the numerous audit and oversight bodies with jurisdiction to investigate DOD contracts may dissuade some companies from competing for DOD contracts. These oversight bodies include the Defense Contract Audit Agency (DCAA), Defense Contract Management Agency (DCMA), GAO, Inspectors General (including in some cases the Special Inspector General for Iraq Reconstruction and the Special Inspector General for Afghanistan Reconstruction), and Congress itself in the form of hearings. On December 13, 2011, the Panel met with the Director of the DCAA, Patrick Fitzgerald and the Director of DCMA, Charlie Williams for a briefing to discuss challenges within the contracting community in the DOD. Over the past two decades, both the DCAA and DCMA have substantially decreased staffing while DOD spending and contracting increased exponentially. DCAA's staffing has decreased by approximately 40% since 1990 while workload has increased approximately 140%. GAO officials reported in recent congressional testimony that DCMA's workforce decreased from an estimate of about 24,000 in 1990 to a low of about 9,300 in 2008, that a rebuilding effort was underway but may require increased funding to sustain.[187] Both Mr. Fitzgerald and Mr. Williams emphasized the need for regrowing their workforces to meet the demand within the Department and are making strides to do so.

One method of checking the effectiveness of DOD audit organizations is through a peer review process. According to the Department of Defense Inspector General, DCAA has not had an audit organization peer review in approximately five years. The DLA audit organization failed its peer review and some observers are concerned that defense audit organizations will also fail peer reviews.

Intellectual Property Rights

Federal contracts generally—and DOD contracts particularly—give the government broad rights vis-à-vis the two types of intellectual property that are arguably most likely to be of concern to small and midsize businesses: (1) patent rights,[188] and (2) rights in technical data.[189] Smaller businesses can experience particular difficulties in protecting their rights because of their size and the comparatively limited resources available to them.

Patent Rights

Federal law gives the government broad rights to use patented inventions, although the basis for and nature of these rights depends upon whether the patentable subject is discovered during performance of a government contract.[190] Federal contracts provide that, when the invention was made in the performance of work under a government contract or subcontract, the government "shall have at least a nonexclusive, nontransferable, irrevocable, paid-up license to practice, or have practiced for or on behalf of the United States, any subject invention throughout the world."[191] This license authorizes the government to allow future contractors to manufacture or use the patented invention "for or on behalf of" the government at any time during the term of the patent's protections without obtaining a license from or compensating the patent holder. However, even when the invention was not made in the performance of work under a government contract or subcontract, the government may delegate its power of "eminent domain" over patents to its contractors by including an "Authorization and Consent" clause in the contract.[192] The standard version of this clause provides that:

> The Government authorizes and consents to all use and manufacture, in performing this contract or any subcontract at any tier, of any invention described in and covered by a United States patent—(1) [e]mbodied in the structure or composition of any article the delivery of which is accepted by the Government under this contract; or (2) [u]sed in machinery, tools, or methods whose use necessarily results from compliance by the Contractor or a subcontractor with (i) specifications or written provisions forming a part of this contract or (ii) specific written instructions given by the Contracting Officer directing the manner of performance.[193]

The standard "Authorization and Consent" clause also provides that "the Government assumes liability for ... infringement to the extent of the authorization and consent hereinabove granted," which generally means that the patent owner must sue the government—not the contractor acting on the government's behalf—in the U.S. Court of Federal Claims to recover for the unauthorized use of its inventions.[194] Allegedly "broad" use of this clause has been cited as a particular concern by DOD contractors and subcontractors, who note that suing the federal government in the Court of Federal Claims involves a "lengthy and costly" procedure that limits the ability of small businesses, in particular, "to recover fair and reasonable compensation for their property."[195]

Rights in Technical Data

Federal law gives DOD similarly broad rights in technical data. These rights are broadest when noncommercial items are involved,[196] with the government acquiring either "unlimited rights," "government purpose rights," or "limited rights" in the data.[197] The type and extent of the government's rights depends upon whether the government wholly or partially funded development of the item, component or process to which the data pertains, among other things. However, even when commercial items are involved, DOD may still require the contractor to "[r]elinquish to, or otherwise provide, the Government rights to use, modify, reproduce, release, perform, display, or disclose" certain technical data.[198] This includes technical data that: (1) are "form, fit, or function data;"[199] (2) are required for the repair or operation of commercial items or processes, or for the proper installation, operating, or handling of commercial items, either as standalone units or as parts of military systems; or (3) describe modifications made at government expense to commercial items or processes in order to meet the requirements of a government solicitation.[200]

However, while giving the government certain rights federal law also imposes certain restrictions upon the use or disclosure of technical data in which the government has government purpose rights or limited rights. For example, during the "government purpose rights period," the government may not use (or authorize another to use) technical data marked with government purpose rights legends for commercial purposes, or disclose or release data to any person (or authorize others to do so) unless certain conditions are met.[201] DOD contractors, including small businesses, have objected both to the breadth of the rights in technical data that the government acquires under government contracts and subcontracts,[202] and to the government's compliance with the restrictions upon the use, disclosure, or release of technical data in which the government has government purpose or limited rights. Small business contractors, in particular, have alleged that government employees improperly furnished materials to their competitors so that their competitors could "reverse engineer" their proprietary products, as well as undertook research projects duplicating proprietary solutions and then published the resulting intellectual property as government-owned.[203] In fact, in one notable instance, the Court of Federal Claims took the unusual step of awarding the contractor expectation damages for lost profits after the government "repeatedly breached the [Cooperative Research and Development Agreement] by releasing the plaintiff's proprietary information to unauthorized recipients, including its competitors."[204]

Intellectual Property Rights under DOD SBIR Contracts Specifically

When technical data or computer software or software documentation is to be generated during performance of contracts under the SBIR program,[205] the contract provides greater protections for the contractor's intellectual property than DOD contracts typically do. Under the standard "Rights in Noncommercial Technical Data and Computer Software—Small Business Innovation Research Program" clause, defense agencies obtain only a "royalty-free license to use data marked with an SBIR data rights legend... for government purposes during the period commencing with contract award and ending five years after completion of the project under which the data were generated."[206] Moreover, federal regulations expressly prohibit the government from releasing or disclosing SBIR data during the license period to any person other than its support services contractors except (1) for evaluation purposes, (2) as expressly permitted by the contractor, or (3) when use, release, or disclosure is necessary for emergency repair or overhaul of items operated by the government.[207] In addition, a 2002 SBA Policy Directive prohibited agencies from using information from SBIR programs to produce technical procurement specifications that could harm the SBIR company that discovered and developed the innovation.[208]

SBIR contractors, however, have raised serious concerns about the degree to which DOD complies with the restrictions in existing law and policy.[209] In particular, contractors have reported difficulties in retaining their data rights in Phase III of the SBIR program, which involves commercialization of the project. Many report that agencies and/or prime contractors "pressure" them to turn over their rights, or "fight them" in their attempts to retain their rights.[210] Some firms also allege that agencies improperly procure goods through non-SBIR contracts that are follow-on to Phase II SBIR contracts, and should go to SBIR firms.[211]

Export Control

> "If there is anything more mysterious than FAR and DFARS, it is ITAR. And if there is anything that small companies know less about than FAR and DFARS, it is ITAR."[212]
>
> Mr. Joel L. Johnson
> Former Vice President, International
> Aerospace Industries Association of America, Inc.

For years, business has complained about what it considers the cumbersome and restrictive export licensing system of the Department of State's Directorate of Defense Trade Controls (DDTC). Only recently has the Defense Security Service (DSS) become more forward leaning in working with business to understand the threat the private sector faces through the theft and misuse of U.S. defense technology.

Some restrictions in the current system may disproportionately affect small business in that they may not have the corporate resources to navigate the often complex export licensing system and the result can be the elimination in the world market (e.g. if a commercial firm's part is also contained in military system).[213] Some of the impediments pointed out by business include:

- **Registration.** All manufacturers and exporters of defense articles or furnishers of defense services must register annually with DDTC whether they export or not during the year. The registration fee is $2,250 for new applicants or those who have not exported in the previous year, $2,750 for registrants who have had between 1-10 licenses reviewed by DDTC during the previous year, and $2,750 plus $250 times the number of licenses over 10 reviewed the previous year.
- **Licensing Requirements.** Items designated as defense articles on the U.S. Munitions Lists, or judged to be defense articles through a commodity classification process, including parts and components for that item, are subject to the licensing requirements of ITAR. Except for certain exceptions to Canada, Australia, and the United Kingdom, and certain other license exemptions, a license is required for all transactions. In addition, a presumption of denial exists for licenses to 18 countries due to U.S. or international arms embargoes, including the People's Republic of China.

- **Technical Assistance Agreements and Manufacturing Licensing Agreements.** An agreement for the performance of defense services or the disclosure of technical data or an agreement granting a right or license to manufacture defense articles abroad must be approved in advance by DDTC.

Export Control Reform Efforts

In the current export control system, responsibility for controlling exports is divided among the Commerce, State, and Treasury Departments based on the nature of the product (munitions or dual-use goods) and basis for control, with enforcement shared among these agencies as well as the Departments of Justice and Homeland Security. During the 111th Congress, the Obama Administration announced the launch of a comprehensive review of the U.S. export control system. Defense Secretary Robert M. Gates announced key elements of the Administration's agenda for reform in a speech on April 20, 2010, with additional elaborations in subsequent months. Secretary Gates proposed a four-pronged approach that would create a single export control licensing agency for both dual-use and munitions exports, adopt a unified control list, create a single integrated information technology system, which would include a single database of sanctioned and denied parties, and establish a single enforcement coordination agency. While this reform effort would likely not fundamentally change these defense trade licensing and registration requirements, it has the potential to ameliorate some of the concerns of the business community primarily through paring the U.S. Munitions List (USML). At the same time, such paring has the potential to allow sensitive U.S. technologies to be obtained more readily by adversaries, The Senate Select Committee on Intelligence, in its report accompanying S. 1458 for fiscal year 2012 Intelligence Authorization Act, required a review by the Intelligence Community (IC), due in mid-March, to evaluate threats to U.S. security by "technological export" and the role of the intelligence community in the Administration's export control review process.[214] In addition, other reforms contemplated, such as the adoption of a single-IT system, may have the potential to reduce licensing times in cases where a defense license is still required.

Legislative Actions

Three bills have been introduced in the 112th Congress related to export controls. The Export Administration Act (EAA) Renewal Act of 2012 (H.R. 2122) would renew the 1979 EAA until 2015, provide enhanced penalty and enforcement authority, provide for congressional review of export control regulations, toughen Iran sanctions, and authorize differential treatment of parts and components on the USML. Meanwhile, the Technology Security Act of 2011 (H.R.2004, Berman) would completely rewrite the EAA, vesting the President with the authority to control exports for national security, foreign policy, proliferation, terrorism, or disruption of critical infrastructure reasons under certain guidelines. A third bill, the Safeguarding United States Satellite Leadership and Security Act of 2011 (H.R.3288, Berman) would authorize the President to remove commercial communications satellites and related components from the USML to the less-restrictive Commerce Control List (CCL). It is worth noting that the bipartisan Cox Commission, chaired by former Representative Christopher Cox, specifically recommended the regulation of commercial satellites and related components through the USML following a lengthy bipartisan review and a unanimous report concluding that the People's Republic of China had been able to obtain U.S. technology that materially improved Chinese ballistic missile capability when that technology was controlled by the CCL.[215]

It is imperative that we find the right balance between protecting our military advantage and growing our industrial competitiveness. Noted in the appendices containing the summary of the Panel Roundtable discussions, several small defense suppliers described how dated or overly restrictive controls are negatively affecting their domestic competitiveness and global market share.

Domestic Sourcing Requirements

Sometimes, the laws and regulations governing defense procurement can add to the costs of doing business, as may sometimes occur in the case of certain domestic source restrictions like the Berry Amendment and the Buy America Act.

The Buy American Act

Enacted in 1933, the Buy American Act governs domestic preference in procurement of the federal government. The Buy American Act establishes domestic preference requirement for "articles, materials, and supplies" when they are acquired for public use unless a specific exemption applies. The Act applies to all federal procurements, but has separate provisions for supply contracts and construction contracts. The Act does not apply to procurements to which application would be inconsistent with the public interest or unreasonable in cost. Additionally, the Act does not apply to procurements of products for use outside the United States or of products not produced or manufactured in the United States in sufficient and reasonably available commercial quantities and of satisfactory quality. Lastly, the Act does not apply to procurements under $3,000.[216]

The Berry Amendment

Congress and DOD have long debated the need to protect the U.S. DIB by restricting certain federal procurement to U.S. markets through legislation known as "domestic source restrictions." Many defense appropriations bills passed since 1942 have included some mention of a preference for U.S. articles, supplies, and materials. One particular group of domestic source restrictions was first enacted into law on April 5, 1941, as part of the fiscal year 1941 Fifth Supplemental National Defense Appropriations Act, P.L. 77-29. This legislation would come to be known as the Berry Amendment. On December 13, 2001, the passage of the fiscal year 2002 NDAA codified and modified the Berry Amendment, making it a permanent part of the United States Code. The Berry Amendment prohibits DOD from using appropriated funds to purchase items that have not been grown, processed, or manufactured in the United States. Under the Berry Amendment, the Secretary of Defense has the authority to waive the domestic source requirement, under certain conditions.[217]

The Berry Amendment, which dates from the eve of World War II, was established for a narrowly defined purpose: to ensure that U.S. troops wore military uniforms wholly produced within the United States and that U.S. troops were fed with food products entirely produced in the United States. Other industries were added later.

The Jones Act

The Jones Act, established by section 27 of the Merchant Marine Act of 1920 (46 U.S.C 883), requires that all waterborne shipping between points in the United States be carried by vessels built in the United States and operated by Americans. The purpose of the Act is to ensure that the nation has a sufficient merchant marine and shipbuilding base to protect the nation's defense and commercial interests. Critics claim that the Act does not accomplish this goal and furthermore raises shipping costs, thereby making U.S. farmers and manufacturers less competitive. Jones Act supporters claim that the Act is needed to foster a domestic shipbuilding base that is vital for national security. Despite economic arguments against the Jones Act, efforts to repeal the Act have not been successful.[218]

Challenges to the Application and Enforcement of Domestic Sourcing Requirements

Some critics of the these policies have argued that the restrictions may not always represent the best value to DOD or the federal government, nor is there always a justifiable national security interest to preserve certain items under domestic sourcing requirements. For example, compliance with the Berry Amendment may make it more difficult for DOD to take advantage of commercial business practices. In an increasingly globalized economy, a few suppliers find it difficult to adhere to these restrictions as they often deviate from standard commercial business practices, thus some suppliers may decline to sell to DOD. Some suppliers who sell to DOD claim they are often forced to adopt unique, costly, and inefficient business practices to do business in the defense sector.

On the other hand, some advocates of domestic sourcing policies have asserted that U.S. workers and businesses have an expectation that Congress will consider their interests in preserving a strong domestic industrial base. They argue that policies like the Berry Amendment mean that the United States will become less dependent on foreign sources of supply, and assert that if the United States becomes dependent on purchasing equipment and supplies from foreign sources, what would prevent an adversary from cutting off U.S. access to such items or refusing to build militarily critical items in times of crisis or conflict? Another argument for maintaining these restrictions is that they can often benefit small, minority-owned, veteran-owned, women-owned, and other types of small businesses which may depend on DOD for their viability.

Additional Costs and Risks Associated with Doing Business with DOD

As discussed, government contracting is different than private sector contracting. Many analysts and defense businesses argue that another challenge to doing business with DOD is the instability of certain government contracts. For example, many DOD contracts have a base year and follow-on option years. The general lack of long-term contracts may create business uncertainty for contractors. In addition, budget uncertainty, frequent changes to the funding and quantity buy of some programs, and multiple award contracts, may all inject uncertainty into the business planning of companies seeking to work for the DOD.[219] A GAO report stated, "in its December 1994 report, *The DOD Regulatory Cost Premium: A Quantitative Assessment*, Coopers and Lybrand identified over 120 regulatory and statutory "cost drivers" that, according to the contractors surveyed, increase the price DOD pays for goods and services by 18 percent." 220 Despite the many acquisition reform efforts have taken place since that time, it is likely that costs, due to added regulations, have only increased.

Raytheon Corporation's public filing with the Securities and Exchange Commission for fiscal year 2011 reflects how some contractors view these unique risks associated with doing business with the Department of Defense. In the filing, Raytheon states:

> Government contractors must [also] comply with specific procurement regulations and other requirements. These requirements, although customary in government contracts, impact our performance and compliance costs. In addition, current U.S. Government budgetary constraints could lead to changes in the procurement environment, including the DOD's recent initiative focused on efficiencies, affordability and cost growth and other changes to its procurement practices. If and to the extent such changes occur, they could impact our results of operations and liquidity...
>
> In addition, failure to comply with these regulations and requirements could result in reductions of the value of contracts, contract modifications or termination, and the assessment of penalties and fines, which could negatively impact our results of operations and financial condition...
>
> As a government contractor, we are subject to routine audits and investigations by U.S. Government agencies such as the Defense Contract Audit Agency (DCAA). These agencies review a contractor's performance under its contracts, cost structure and compliance with applicable laws, regulations and standards. The DCAA also reviews the adequacy of and a contractor's compliance with its internal control systems and policies, including the contractor's purchasing, property, estimating, compensation and management information systems.[221]

Acquisition Reform Efforts

There is much debate over how effective the numerous acquisition reform efforts have been. Compared to the era of the Civil War, when the government

bought weapons that did not work, horses that were diseased, and food that was rotten, one can argue that the process has improved. However, it is uncertain whether reform efforts of the last fifty years have generally accomplished their aims. Some analysts have argued that acquisition reform efforts have made the process less efficient and effective, prompting calls to stop embarking on successive rounds of acquisition reform.[222] Others have argued that the fundamental problem with DOD acquisitions is not policy, but execution and expectations.[223] Some of these analysts argue that the way to improve defense acquisitions is to improve the quality, quantity, and incentive structure of the acquisition workforce.

One area of concern where clear progress has not been visible is the cost growth of Major Defense Acquisition Systems. In the early 1980s, a number of major weapons systems programs were experiencing dramatic cost overruns, overruns which increased the defense budget by billions of dollars but resulted in the same number, or in some cases fewer, weapons. Programs experiencing cost growth included the Patriot missile system (37% cost growth), the Hellfire missile (48% growth), the Blackhawk helicopter (24% growth), and the F-18 (21% growth). According to the December 1980 Selected Acquisition Report, there was a $47 billion cost increase for 47 major weapon systems in just the last three months of 1980.[224]

According to many analysts, since the 1970s and 1980s, the extent of cost growth of weapons systems has generally remained the same.[225] As one RAND report stated, "cost growth has not improved over the decades.... Thus, despite the many acquisition reform and other DOD management initiatives over the years, the acquisition cost growth of military systems has not been reduced."[226]

As noted, acquisition reform efforts have historically focused more on reducing cost to develop and procure a weapon system and less on the actual total ownership costs of a weapon system. As one example, the Nunn-McCurdy Act (10 U.S.C. §2433), which was included in the Department of Defense Authorization Act, 1983 (P.L. 97-252), requires DOD to report to Congress whenever a major defense acquisition program experiences cost overruns that exceed certain thresholds. One of the goals of the Act was to help control cost growth in major defense systems by holding the appropriate Pentagon officials and defense contractors publicly accountable and responsible for managing procurement costs.[227] Because the Act focuses on procurement costs and not total ownership costs, it may be driving behaviors and decision making during development that end up increasing total ownership costs. Yet operation and sustainment costs generally account for

70% or more of overall weapon system costs. There is no similar requirement to notify Congress when the operation and sustainment costs of a program exceed cost estimates.

One example where efforts to reduce procurement costs may have led to unnecessary and expensive sustainment efforts is corrosion on the F-22 fighter aircraft. According to GAO, corrosion was found in the F-22 fleet in early 2005, less than 6 months after the aircraft were fielded in a coastal environment. By October 2007, a total of 534 instances of corrosion were documented and at, that time, the government estimated it would cost over $228 million to make F-22 corrosion-related repairs and retrofits through 2016.[228] GAO stated that no operational-level test for corrosion was conducted on the F-22 prior to initial operating capability, and the length of the F-22 full-scale climatic test was cut in half. At Congressional direction, DOD reviewed the matter and concluded that, "if the F-22 program had accomplished testing earlier in the program, many of the corrosion problems could have been addressed at greatly reduced cost and the associated readiness issues avoided."[229]

The Weapon Systems Acquisition Reform Act of 2009

One of the more recent efforts to reform weapon system acquisitions was the Weapon Systems Acquisition Reform Act of 2009. Key provisions in the act included:

1. The appointment of a Director of Cost Assessment and Program Evaluation;
2. The appointment of a Director of Developmental Test and Evaluation;
3. The appointment of a Director of Systems Engineering;
4. A requirement that the Director of Defense Research and Engineering periodically assess technological maturity of major defense acquisition programs and annually report finding to Congress; and
5. A requirement that combatant commanders have more influence in the requirements generation process.

The extent to which these reforms are successful will depend in part on whether DOD fully embraces and supports these initiatives. Given how recently the Weapon System Reform Act was enacted, only time will tell to what extent this effort will be more effective than past efforts to improve the performance of major defense acquisition programs.

Recommendations

As discussed above, a number of hurdles make it challenging for companies to compete for defense contracts. The plethora of regulations specific to government and defense contracting dissuades many companies from competing for government contracts. The acquisition process is often bureaucratic and rigid, with insufficient flexibility to allow appropriate application of management, oversight, and monitoring of small businesses. The defense business environment is also complicated, and some argue hindered, by current export control requirements. The high rate of personnel turnover in government acquisition personnel, from program managers to DCAA auditors affects the quality and consistency of policies. Oversight and management agencies such as DCAA and DCMA are under-resourced and lack consistently trained, skilled personnel, hampering the ability of these agencies to provide appropriate contract oversight and management. In addition, a backlog of audits has caused DCAA to prioritize work on high dollar contracts, leaving unresolved many of the open audits of small businesses who are holding small dollar contracts. In light of these and other concerns, the Panel recommends:

RECOMMENDATION 5.1: Congress should require the DOD IG to notify Congress in its semi-annual report if a DOD audit organization has not complete a peer review within the standard three year period or if a DOD audit organization fails a peer review.

RECOMMENDATION 5.2: Congress should require the Council of Inspectors General on Integrity and Efficiency to report to Congress whenever a member of the Council does not complete a peer review within the standard three year period or whenever a member audit organization fails a peer review. Any newly created member of the council should be exempt from reporting on whether it completed a peer review for the first five years from the date of the creation of the member organization.

RECOMMENDATION 5.3: The Secretary of Defense and Congress should undertake a comprehensive review of the laws and regulations that govern the acquisition process in an effort to repeal or amend regulations that are outdated or have had unintended consequences that outweigh the original intent of the regulation. This effort should be undertaken with an eye to simplifying and streamlining all aspects of the acquisition process and reducing the negative cost and schedule impacts.

RECOMMENDATION 5.4: The DOD IG should assess the degree to which DOD complies with the restrictions in existing law and policy related to use, disclosure, or release of intellectual property, with specific emphasis on intellectual property developed under the SBIR program.

RECOMMENDATION 5.5: Congress should direct the Secretary of Defense to increase oversight of the management, functionality, and operations of DCAA and DCMA to reduce the backlog of audits, and to improve the audit agencies' relationship with the industrial base. The Panel is aware that DCAA executives have met with a variety of industry associations over the last few years to further ensure that they have effective communication with their contracting community. These meetings and other forms of engagement with industry should be continued into the future.

RECOMMENDATION 5.6: Congress should direct the Secretary of Defense to examine the Department's organizational structure and assess the feasibility and advisability of reorganizing the Department to realign DCAA and DCMA to improve communications, audit performance, oversight, and management.

RECOMMENDATION 5.7: Congress should direct the Secretary of Defense to establish a small business advocacy office and a contract close out unit in DCAA and DCMA to ensure that the needs of small businesses are safeguarded and that all contracts are closed out in a timely fashion. Closing out contracts in a timely fashion is a key element in having auditable financial statements.

RECOMMENDATION 5.8: The Directors of DCMA and DCAA should ensure coordination between their agencies and the SBA when conducting audits that include factors of interest to or duplicative of reviews conducted by SBA. For example, SBA's Commercial Market Representatives visit large contractors with subcontracting plans to assess compliance with the subcontracting plan. However, DCAA also looks at subcontracting as part of its cost audits, especially when subcontracting performance is related to a company's award fee. Furthermore, DCMA also reviews subcontracting performance and processes. These three entities should coordinate their reviews to more efficiently conduct audits and to potential reduce the number of audits performed.

RECOMMENDATION 5.9: Congress should examine other alternatives, to include the establishment of a self-regulatory option, to providing auditing, accounting and advisory services regarding contracts and subcontracts and examine the feasibility of using such alternatives for the DIB to potentially reduce or eliminate many of DOD's internal audit organizations while ensuring compliance with statutory, regulatory, and contractual requirements.

RECOMMENDATION 5.10: Congress should require the Secretaries of State and Defense report to Congress their goals for speeding the review of defense article export licensing applications and their proposals for administrative changes needed to reach those goals.

RECOMMENDATION 5.11: Congress and the Administration should work together to conduct a comprehensive reform effort of U.S. export controls. The current system is overly complex, subjectively enforced and unable to respond to advances in the global markets and technology developments. Independent of a comprehensive review, a process under subsection 38f of the Arms Export Control Act to move USML controlled items to the less rigorous CCL in consultation with the appropriate congressional committees exists.[230] This process, with appropriate consultation, should continue to be utilized.

APPENDIX A – PANEL ON BUSINESS CHALLENGES IN THE DEFENSE INDUSTRY WORK PLAN

Panel Members

Bill Shuster, PA, Chairman
Bobby Schilling, IL
Jon Runyan, NJ
Allen West, FL

Rick Larsen, WA, Ranking Member
Betty Sutton, OH
Colleen Hanabusa, HI

Rules and Procedures

The Panel is constituted under Rule 5(a) of the rules of the Committee on Armed Services to serve for a period of six months beginning on the date of its organization, September 12, 2011. The Chairman of the Committee has the discretion to reappoint the Panel for a period of time that may extend to an additional six months. The Panel will follow the rules and procedures of the

House Committee on Armed Services, as adopted by the Committee for the 112th Congress, in all of its meetings, hearings, and other activities. The Panel has been tasked to examine the current defense business environment and to seek to understand how the Department of Defense (DOD) could spur innovation, competition, and cost savings by encouraging new entrants into the industrial base and by fostering the transition of technology. Although it does not have legislative jurisdiction, the Panel will report its findings, including any recommendations for possible legislation, to the Committee.

Staffing

The Panel will be assisted by staff of the House Committee on Armed Services designated by the Chairman and Ranking Member of the Committee for this purpose.

Work Plan

The Panel will examine the current defense business environment to identify (1) contracting or regulatory issues facing the defense industry; (2) the use of incentives and mandates to meet goals; (3) structural challenges facing various sectors within of the industrial base, including universities and research institutes; (4) impact of the current fiscal environment on the health of the defense industry, at both the prime and subcontractor levels; and (5) opportunities to reduce barriers to entry. In examining business challenges within the defense industry the Panel will focus on five primary areas:

- The use of incentives and mandates to shape the defense business environment
    - Are there incentives to be entrepreneurial and/or innovative?
    - What is the intent of mandatory set-asides? Are mandates having the desired effect? Does the Department meet the requirements/goals of these programs? How does verification work at lower subcontract tiers?
    - Do set-aside programs disincentivize growth because companies don't want to grow out of being "small"?
    - How does contract bundling reduce competition?
- Barriers to transitioning technology

- What challenges do smaller firms have getting beyond R&D and into test and integration?
- Does DOD aversion to risk hamper initiative and innovation in industry?
- Do financial limitations hamper industry's ability to transition technology?
- Are non-traditional firms marginalized in the business industry, and if so, what are the causes?

- Impact of regulation/standard contract clauses and associated enforcement on defense businesses
    - Is there sufficient flexibility to allow appropriate application on small business?
    - Is there lack of awareness/training in such application for DOD and contract oversight agencies?
    - Are requirements too burdensome to attract new entrants?
- What is the role of universities and government laboratories in terms of spin outs, technology transition, and partnering with small businesses?
- Economic projections and access to capital for defense businesses.

In addition to these matters, the Panel may examine other issues related to business challenges within the defense industry at the request of the Committee's Chairman with the concurrence of the Committee's Ranking Member.

APPENDIX B – PANEL EVENTS

DATE	TITLE	WITNESSES
September 12, 2011	Member Business Meeting	N/A
September 20, 2011	Hearing on "Challenges to Doing Business with the Department of Defense"	Mr. A.R. Hodgkins, III Mr. Bradford L. Smith, Jr. Ms. Heidi Jacobus
October 7, 2011	Industry Roundtable: Rock Island, Illinois	See Appendix C for Participants
October 14, 2011	Briefing from the Defense Security Service, Counterintelligence Directorate	Mr. Bill Stephens

(Continued)

DATE	TITLE	WITNESSES
October 24, 2011	Hearing on "The Defense Industrial Base: A National Security Imperative"	Mr. Barry Watts Mr. Fred Downey Mr. Pierre Chao
October 28, 2011	Industry Roundtable: Akron, Ohio	See Appendix D for Participants
November 1, 2011	Hearing on "The Defense Industrial Base: The Role of the Department of Defense"	Mr. Brett B. Lambert Mr. André Gudger
November 18, 2011	Hearing on "Creating a Defense Industrial Base for the 21st Century"	Honorable Jacques S. Gansler Mr. David Berteau
December 9, 2011	Industry Roundtable: Mount Laurel, New Jersey	See Appendix E for Participants
December 13, 2011	Briefing on the Defense Contract Management Agency and the Defense Contract Audit Agency	Mr. Charlie E. Williams, Jr. Mr. Patrick Fitzgerald
January 9, 2012	Industry Roundtable: Santa Clarita, California	See Appendix F for Participants
January 10, 2012	Industry Roundtable: Honolulu, Hawaii	See Appendix G for Participants
January 12, 2012	Industry Roundtable: San Diego, California	See Appendix H for Participants
January 17, 2012	Hearing on "Doing Business with DOD: Unique Challenges Faced by Small and Midsize Businesses"	Mr. A. John Shoraka Ms. Linda Hillmer Ms. Lynn Schubert
January 23, 2012	Hearing on "Doing Business with DOD: Getting Innovative Solutions from Concept to the Hands of the Warfighter"	Dr. Stephen E. Cross Dr. Norman Winarsky Dr. Stephen Huffman
February 6, 2012	Hearing on "Doing Business with DOD: Contracting and Regulatory Challenges"	Dr. Allan Burman Mr. Raj Sharma* Mr. Joel Johnson
February 21, 2012	Industry Roundtable: Palm Beach, Florida	See Appendix I for Participants
February 27, 2012	Industry Roundtable: Washington, DC	See Appendix J for Participants

*Provided written testimony but did not appear due to illness.

APPENDIX C – SUMMARY OF INDUSTRY ROUNDTABLE AT ROCK ISLAND, ILLINOIS

Overview

On October 7, 2011, Members of the Panel of Business Challenges in the Defense Industry traveled to the Quad Cities area of Illinois and Iowa to meet with members of the local defense industry. The Quad Cities area is home to Rock Island Arsenal (RIA) which comprises 946 acres on the Mississippi River between the cities of Davenport, IA, and Rock Island, IL. The island was originally established as a government site in 1816, with the building of Fort Armstrong. It is now the largest government-owned weapons manufacturing arsenal in the United States. It has been an active manufacturer of military equipment and ordnance since the 1880s. Established as both an arsenal and a center for the manufacture of leather accoutrements and field gear, today RIA provides manufacturing, logistics, and base support services for the Armed Forces. The Arsenal is the only active U.S. Army foundry, and manufactures ordnance and equipment, including artillery, gun mounts, recoil mechanisms, small arms, aircraft weapons sub-systems, grenade launchers, weapons simulators, and a host of associated components. Some of the Arsenal's most successful products include the M198 and M119 towed howitzers, and the M1A1 gun mount. Approximately 250 military personnel and 6,000 civilians work at the arsenal.

The Panel held a roundtable discussion with representatives from the defense industry at Black Hawk College, Moline, IL. Following the discussion the delegation met with RIA leadership, and toured the Joint Manufacturing and Technology Center, the primary tenant on Rock Island, and other Arsenal facilities.

Members in Attendance

Chairman Shuster
Ranking Member Larsen
Mr. Schilling Ms. Hanabusa Mr. Loebsack

Points of Discussion

- **Defense Contract Audit Agency** – Several participants indicated that continued and significant backlogs at DCAA and the complexity of incurred cost audits tend to hold up contract closeouts. A comment was made that one person was needed full time simply to support compliance/audit requirements and that it was difficult to self-audit.
- **Achieving Small Business Goals** – Many in industry felt that DOD was not achieving its Small Business goals (see figure 1) and that DOD contracting officials were not doing enough to hold prime contractors responsible for meeting the Small Business Contracting Plans submitted as part of the bid and proposal processes. Other comments:
- Past performance (quality) regarding Small Business participation is not being considered in source selection
- DOD's tendency toward seeking "lowest price/technically acceptable" solutions is bad for small business because small businesses cannot "buy-in" [achieve economic order quantities] to reduce unit costs the way large companies can. Low price consideration over quality also generally means inferior product to the Warfighter.
- Some form of protection for businesses that have graduated from being a small business would be helpful because mid-tier businesses who no longer qualify for small business assistance cannot compete against the larger defense firms.
- Work is not being passed through to small businesses in accordance with their proposed Small Business Contracting Plans. An example was given that on a $1 billion contract, only $35 million was subcontracted to small business. Many felt that there is a need for better accountability for meeting small business goals.

Prime Contracting Goals	FY 2009	FY 2010	FY 2011
Small Business	22.24%	22.28%	22.28%
HUBZone Small Business	3.0%	3.0%	3.0%
Service-Disabled Veteran-Owned Small Business	3.0%	3.0%	3.0%
Small Disadvantaged Business *	5.0%	5.0%	5.0%
Women-Owned Small Business	5.0%	5.0%	5.0%
Historically Black Colleges & Universities and Minority Institutions **	5.0%		

Subcontracting Goals	FY 2009	FY 2010	FY 2011
Small Business	37.2%	31.7%	31.7%
HUBZone Small Business	3.0%	3.0%	3.0%
Service-Disabled Veteran-Owned Small Business	3.0%	3.0%	3.0%
Small Disadvantaged Business	5.0%	5.0%	5.0%
Women-Owned Small Business	5.0%	5.0%	5.0%
Historically Black Colleges & Universities and Minority Institutions ***	5.0%		

* Small disadvantaged business awards include 8(a) awards.

** Base for HBCU/MI measurement is awards to Higher Educational Institutions.

***Defense Components are not required by DOD to establish separate HBCU/MI subcontracting goals. Instead these awards should be included when developing the subcontracting goals for Small Disadvantaged Business.

Figure 1. DOD Prime and Subcontracting Goals.

- **Industry Needs Clear Demand Signals** – One participant stated that small businesses are very good at being innovative and responsive, but DOD needs to do a better job of providing a clear demand signal and allowing industry sufficient time to respond. An example was given regarding a DOD requirement for 200 obsolete grinders and industry was only given 3 days to prepare a proposal in response. [Members of the Panel commented that some of this behavior may be caused by Congress failing to provide a full year appropriation at the start of the fiscal year.]
- **Flexibility in Cost/Pricing** – Comments were made that lengthy contracting periods (3-5 years) on firm-fixed price contracts require industry to “guess” on costs to procure raw materials for the period of performance. While larger industry may be able to absorb price fluctuations, many small business do not have the cash flow to do so. Industry participants indicated that commercial sales contracts use a price index and/or review at the 2- and 4-year points to address this issue and suggest that DOD and the defense industrial base would benefit from adopting that model.
- **Small Business Innovative Research Program** – there was strong concurrence that the SBIR program is very difficult to use and needs to be simplified. Of the participants in the room, few had actually participated in or benefitted from the SBIR program. Note: since most the effort at RIA involves manufacturing, it is not unusual for there to be a low number of participants in a program designed to foster innovation.

- **Prime Contractors in Control** – some participants indicated that big business was predisposed to an engineering solution and a small set of subcontractors that they consistently use. It was stated that it was very difficult to get the primes to listen to small business ideas and that "big business is in control." It was suggested that small businesses need direct access to the DOD decision-makers/acquisition officials because big business did not want direct competition from small business. One participant specifically stated that he felt that the end-user, not the prime contractor, should make the decision to go with a specific sub-contractor/small business to buy a component/end-item.
- **Funding to Test Developmental Products** – participants indicated that small businesses often lack funding to carry a product from development through testing.
- **Partnering with the Defense Organic Industrial Facilities** – there was strong sentiment that partnering with defense facilities such as arsenals and depots was complicated, but works and is good for small business. Section 4544 of Title 10, United States Code, allows Army industrial facilities to partner with private industry in certain circumstances, but the authority limits the number of partnerships and has a sunset of September 30, 2014. Partnering can take several forms but is generally done in an effort to allow private industry access to and use of government owned facilities, workforce and tooling while allowing the government to benefit from capital investments made by industry, the sharing industry best-practices, and access to intellectual property (e.g. technical data packages) without having to purchase the data. Section 325 of the House-passed version of the NDAA of 2012 would remove the limitation and make the authority permanent. There was strong support for this provision amongst the industry participants and RIA government leadership and they urged adoption of the provision. Partnerships such as this help to preserve DOD's organic capability.
- **International Trafficking of Arms Regulations** – ITAR is a set of government regulations that control the export and import of defense-related articles and services on the USML. Many from industry complain that ITAR is overly restrictive, takes a "one-size fits all" approach to determining what is placed on the USML, and does not allow for items to easily be moved off the list as technology advances and specific items or subcomponents become readily available on the global market. One of the industry participants jokingly referenced

ITAR as a "jobs program" because of the time and effort industry spends trying to comply with the regulations.

Industry Roundtable Participants

Alcoa Davenport Works - Mr. Tony Morales
Mayor Bill Gluba - Mayor of Davenport, IA
Group O - Ms. Dottie Tubbs, Vice President of Operations
Honsa Ergonomic Technologies, Inc. - Mr.Thomas Honsa, Chief Executive Officer
Illinois Procurement Technical Assistance Center at Black Hawk Community College – Ms. Vicky Miller, Center Director
Mandus Group - Mr. Sam Kupresin, Vice President
McLaughlin Body Company - Mr. John Mann, President, & Mr. Tom McLaughlin, Chief Executive Officer
PBC Linear - Mr. John Oller, Manufacturing Manager
Pendulum Resources - Mr. Alan Kruse, Program Manager
Quad Cities Chamber of Commerce - Ms. Tara Barney, Chief Executive Officer
Rock Island Arsenal Development Group - Mr. Tim Frye, Site Manager
Sivyer Stee - Mr. Art Gibeaut, President & Chief Executive Officer
SSAB America - Mr. Jim Barber, Regional Sales and Product Development Manager
SupplyCore - Mr.Steve Cotone, Program Manager & Mr. Mike Paul
Mayor John Thodos - Mayor of East Moline, IL
Vista International - Mr. Craig Roberts, President & Chief Executive Officer

APPENDIX D – SUMMARY OF INDUSTRY ROUNDTABLE AT AKRON, OHIO

Overview

On October 28, 2011, Members of the Panel of Business Challenges in the Defense Industry traveled to Akron, OH to meet with members of the local defense industry. The Panel held a roundtable discussion at the University of Akron with representatives from the local defense industry (see Appendix A for a list of participants). The University of Akron is a public, co-ed university and has over 24,400 undergraduate students, 4,650 postgraduate students, and operates with an $133 million endowment. The University is most widely known for its research programs in polymers and advanced materials – both of which have significant implications for the defense industry. In April 2011, the University of Akron partnered with the DOD and the Department of Energy (DOE) to build a $14.8 million research facility. The facility will link Akron faculty and students with the needs of private industry, DOD, and DOE. The facility will house the National Center for Education and Research in

Corrosion and Materials Reliability, which was provided $8 million from DOD.

Following the discussion with local defense industry, the delegation met with University leadership, including the President, Dr. Luis Proenza, as well as officials from Wright-Patterson Air Force Base, the DOD Office of Corrosion Policy and Oversight, and industry (see Appendix B for a list of participants) to discuss research conducted at the University to assist DOD in corrosion prevention and mitigation.

Members in Attendance

Chairman Bill Shuster
Ranking Member Rick Larsen
Ms. Betty Sutton
Ms. Colleen Hanabusa

Points of Discussion

- **Export Controls** – a participant indicated that recent changes in the process for obtaining an export control license are causing delays that are making it difficult for US industries to compete in a global marketplace. It was stated that licensing would previously take one or two weeks and now it routinely takes more than six weeks to obtain a license for the same type product. The participant felt strongly that although foreign/NATO partners would prefer American-made products due to the quality, delays in licensing are driving these customers to contract with non-US manufacturers. Additionally, some foreign sales require that the customer places the money for a sale in an escrow-like account until full delivery of the product is achieved. If delays in the delivery occur for any reason, a penalty is applied. One Akron performer, who has been waiting for more than six weeks for an approval from the DOC is losing money as a result of this type of delay.
- **International Trafficking of Arms Regulations Issues** – a participant expressed frustration that there is no discernment of product in the current implementation of ITAR controls. Furthermore, he felt there seems to be no sense of urgency by those responsible for

making decisions/determinations and many businesses do not have the capital to sustain their program/technology and make it through the process.

- **Defense Contract Audit Agency Issues** – general concern expressed about DCAA processes and conduct of audits:
- One participant cited an example of their company still working with DCAA to try to close out an audit from 2006. He mentioned that the environment with DCAA is very hostile and it appears as if they simply do not want any more suppliers. He went on to reference a 2007 DOD Directive that gave DCAA authority to subpoena information if a contractor failed to respond to a request in less than 3 days. {Note: DODI 7600.2 provides authority for subpoena but does not specify a timeline. It can be reviewed at http://www.dtic.mil/whs /directives/corres/pdf/760002p.pdf. Additionally, a DCAA memo (http://www.dcaa.mil/mmr/08-PAS-042.pdf) dated December 2008 gives guidance to DCAA officials regarding use of subpoena authority. Section 6 of the memo lays out a specific procedure for access to records, and only specifies that a timely due date should be set, and that extenuating circumstances should be taken into account.}
- Another participant felt that DCAA is not satisfied when provided sufficient documentation – they want to know "why" a decision was made even when it was in scope of the contract and fully justifiable. He claimed that if one tried to question DCAA or otherwise disagree with their findings, the auditors threaten to broaden the scope of the audit to other contracts.
- It was also stated that DCAA auditors used to sit down at the table with the business and go over the books, allowing for a dialogue and exchange of information that would not only clear up any audit concerns quickly, but would also help the business to learn the audit process so that they could do better in the future. It was noted that auditors seem to no longer do that and instead appear to be solely on a search for any bit of information that might cast a negative light on the business.
- A participant felt strongly that DCAA needs to exist and plays an important role in oversight, "even as much as we [industry] hate spending 3 days to find $58 from 4 years ago."
- **Overly Burdensome Practices** – one participant believed that DOD was applying a manufacturing mindset to the acquisition processes and stated that in acquisition "better planning does not necessarily

drive better performance". He noted that the pace of business is picking up but DOD contracting is not responding due to increased oversight, mandated disclosures of proprietary information, long contracting periods, etc.

- **Need for Better Access to the Customer** – it was stated that the acquisition process at DOD is extremely complicated and industry has little direct access to the customer to find out what the true requirement is. A participant stated that support contractors (not government R&D personnel) are usually all they [industry] have access to and stovepipes in the DOD R&D portfolio make it very difficult for industry to find out what solutions/technology DOD is interested in. He believes that stove-piping in the R&D portfolio is a result of many different government research entities competing for a limited amount of funding. It was also stated that in order to gain any access to DOD officials, interested businesses are forced to pay high (>$2000) conference fees to attend annual conferences just find out about areas DOD is interested in. Another participant articulated the frustration by saying, "We need to have meetings with the people we're dealing with to remove roadblocks."
- **Changes to Rules of Engagement** – a participant expressed frustration that during the course of a solicitation, DOD will often change the requirements and will provide industry little time to respond. In one case, a company was notified 7 days before the solicitation closed that they would need to be a "cleared facility" {approved for work on classified or sensitive programs by the Defense Security Services}. The clearance process is lengthy and they were unable to compete as a result of the last-minute change.
- **Small Business Innovative Research Program** – comments were made that the SBIR Program is transparent, but no other S&T efforts are visible to industry.
- Regardless of transparency, one participant noted that "SBIR transition is horrible" and claimed that the Air Force gave their data package {resulting from an SBIR effort} to a large business.
- This was followed by a comment that intellectual property rights protections need to be built into the SBIR program.
- Another participant commented that there is {seemingly} no coordination between the DOD Office of Small Business Programs and the SBIR Program.

- A participant felt that Phase III is not clear and that there is no way for small business to secure funding to get through testing. It was felt that limitations on small businesses to complete testing, coupled with DOD's risk aversion, would drive DOD to go to the large primes.
- **Mentor-Protégé Program/Industry Partnering -** many stated that the Mentor-Protégé program worked well, but it needed to be broadened so more could benefit from it. Small businesses may be able to create a technology/solution but many do not have the resources to manufacture and field. It was stated that they need protection of their intellectual property in order to confidently partner with a larger firm that is equipped and resourced to handle production. It was also suggested that some type of a program to provide secured capital to small business to enable them to reach production would be beneficial.
- **Small Business Collaboration** - Another participant noted that most small businesses do not have all the answers, yet they could be successful if they were encouraged to collaborate with other small business. It was suggested that small business collaboration should be allowed to meet the 51% minimum participation for set-aside contracts.
- **Small Business Participation Reporting**- many of the participants felt that DOD needs to do a better job of monitoring small business participation post award and below the first-tier contract. It was suggested that large contractors needed to be incentivized in some way, perhaps financially, to subcontract with small businesses.
- **Contracting with Labs/Research Facilities** – while programs such as the SBIR programs aid in technology development, participants felt that much of their work with labs and research entities such as Defense Advanced Research Projects Agency (DARPA) goes nowhere because there are no customers for a follow-on program. It was noted that DARPA and the labs are not chartered to manufacture an end-product and therefore their focus is on research. As noted by one participant, "They just want to keep studying with no plan to field." There was a feeling that this leaves small businesses at risk because their technology never moves out of the lab.
- **Large Business Influence** – one participant stated that large businesses do not like the entrepreneurial nature of small business and take steps to eliminate competition by buying technology from the small business and then shelving it so it cannot be advanced.

Industry Roundtable Participants

Advance Materials Products, Inc (ADMA) - Georg I. Abakumov, Esq., General Counsel and Director of Business Development.
Armor Source, LLC - Yoav Kapah, President & Chief Product Development Engineer and Paul Garcia, Contracting Officer & Government Liaison
Artisan Industries, Inc. - Mark A. Price, Ownership Group
Fireline TCON, Inc. - Klaus-Markus Peters, General Manager & Director of Engineering, and Rich Lonardo from Defense & Energy Systems
Midwest Industrial Supply - Bob Vitale, Chief Executive Officer
Ohio Aerospace Institute - Don Majcher, Vice President of Technology and Innovation Partnerships
Orbital Research - Fred Lisy, President
Powdermet - Andrew Sherman, President & Chief Executive Officer
Team Wendy LLC - Daniel E. Gibbens, Director of Finance & Contract Administration
Tencate Advanced Armor - Mark Edwards, President
TeraPhysics Corporation - Gerry Mearini, President
Will-Burt - Jeffery Evans, Chairman, Chief Executive Officer & President and Dave Harpley, Head of Government Relations

APPENDIX E – SUMMARY OF INDUSTRY ROUNDTABLE AT MOUNT LAUREL, NEW JERSEY

Overview

On December 9, 2011, Members of the Panel of Business Challenges in the Defense Industry traveled to Mount Laurel, NJ to meet with members of the local defense industry. The Panel held a roundtable discussion at the Burlington County College with representatives from the local defense industry (see Appendix A for a list of participants). Burlington County College was founded in 1966 and is a comprehensive community college serving approximately 14,000 students with facilities Pemberton, Mount Laurel, Mount Holly, Willingboro and also has a presence at Joint Base McGuire-Dix-Lakehurst (JB MDL). Following the discussion with local defense industry, the delegation met with senior officials from the JB MDL and toured facilities to include the Air Force Expeditionary Center, the Battle Lab and static displays of aircraft stationed on the base. JB MDL is home to the 87th Air Base Wing, which provides installation management support for 3,933 facilities with an approximate value of $9.3 billion in physical infrastructure. More than 44,000 Airmen, Soldiers, Sailors, Marines, Coast Guardsmen, civilians and their family members living and working on and around JB MDL contribute to the economic impact for the state of New Jersey.

Members in Attendance

Chairman Bill Shuster
Ranking Member Rick Larsen
Mr. Bobby Schilling
Ms. Colleen Hanabusa
Mr. Jon Runyan
Mr. Frank LoBiondo

Points of Discussion

- **Procurement Technical Assistance Centers** – The Procurement Technical Assistance Program was authorized by Congress in 1985 in an effort to expand the number of businesses capable of participating in the Government Marketplace. Administered by the Department of Defense, Defense Logistics Agency (DLA), the program provides matching funds through cooperative agreements with state and local governments and non-profit organizations for the establishment of Procurement Technical Assistance Centers (PTACs) to provide procurement assistance. A roundtable participant indicated that local PTACs need to be resourced to advertise their services so that companies seeking to do business with the Department of Defense can be made aware of the assistance the local PTACs can provide. More about PTACs can be found at http://www.aptac-us.org/new/index.php.
- **Large Contractors and Impact on the Local Workforce** – a participant indicated that, particularly in base services contracts, the base often contracts with large contractors that are not from the local area. He felt that these contractors often bring in their own workforce and should do more to use local workers.
- **Veteran's Transition Assistance** – a participant advocated for programs such as the "Helmets to Hardhats" program to help service members gain apprenticeship training so that they can successfully enter the workforce. He suggested that there should be some requirement for companies that contract with military facilities to

participant in these programs. You can learn more about "Helmets to Hardhats" at http://helmetstohardhats.org/.

- **Security/Workforce Access to Military Facilities** – one participant indicated that base security processes were an issue for workers performing on contracts for the base. He suggested that there should be a system in place, such as the issuance of standard identification card for contract employees, to facilitate access to the base and reduce wait time for the workforce to process through base security.
- **Technology Development and Transition** – a participant indicated that DOD was doing a lot "right" and stated that of TIME Magazines "50 Best Inventions" of 2011, seven were attributable to defense investment. However, he also noted that "fast track" contracting methods are not working. He felt that the acquisition system is too complex and that a great deal of experience is needed to learn the FAR. As a result, he contends that everyone is just trying to find "workarounds" in the system to get the Warfighter what was needed.
- **Multiple Award and Indefinite Delivery/Indefinite Quantity Contracts** – one of the participants indicated that he believed that DOD's approach to using Multiple Award and Indefinite Delivery/Indefinite Quantity contracts is increasing costs for the government and is detrimental to industry.
- **Contracting Officials** – one of the participants indicated that he felt there was an inherent lack of trust between contracting officials and small businesses. He believed that contracting officials were more inclined to contract with a large, foreign firm than with a local small business because they felt there was risk that the local small business would not perform.

  He felt that more needed to be done to require contracting officials to credit small businesses based on past-performance. Another participant stated that its "hard to get in the door" and that decision makers perceive risk in small businesses that is unfounded.

  He also articulated that Congressionally-directed funding is critical to transitioning technologies developed by small businesses.
- **Defense Logistics Contracting Actions** – a participant indicated that DLA does a poor job of notifying the public (and specifically interested contractors) when solicitations are cancelled.

  He expressed frustration that time, effort and resources were put into a bid proposal and they did not receive notice that the solicitation was cancelled. He also felt that DLA was failing to meet the minimum targets for contracting with Service Disabled Veteran Owned companies.

- **Bundling of Contracts** – a participant stated that DOD contracting officers have no visibility into discussions between prime contractors and subcontractors. He indicated that once a small business was rolled up into a large or bundled contract, contracting officers have no visibility on the fact that large primes pressure subs to cut prices, only to increase profit for the prime. He felt that the government should benefit from cost savings through direct contracting, and should not be bundling contracts.
- **General Services Administration (GSA) Contracting** – one of the participants expressed frustration with GSA contracting processes and encouraged the Panel to also look into those processes.
- **Preference for the Original Equipment Manufacturer (OEM)** – a participant stated that he felt contracting officials had a preference for OEM parts and tend to require parts be provided through a sole-source arrangement with the OEM. He argued that this preference is damaging the industrial base and leaving the nation reliant on foreign suppliers.
- **Mentor-Protégé Programs** – a participant from a large business indicated that more needs to be done to reduce the negative perception of doing business with a small business and in particular, SDVO businesses. It was articulated that the Mentor-Protégé Program is very important and more should be done to expand program.

Industry Roundtable Participants

Digital Systems Group, Inc. - Mr. Joseph T. Mc Carrie, Director of Sales and Marketing

Drexel University - Mr. Brian Keech, Senior Vice President & Executive Director, Office of the President

Dynamic Defense Materials, LLC - Robert A. Lipinski, President and Chief Executive Officer

L-3 Command and Control, Systems, and Software - Mr. John Allen, Director of Business Development and Legislative Affairs

L-3 Communications, Communication System-East (L-3 CSE) – Mr. John Tierney, Vice President of Operations and Strategic Initiatives

Lockheed Martin Corporation Mission Systems and Sensors (MS2): Moorestown NJ operation -Mr. Phil Goslin, Subcontract Program Management Director

MCFA – Mr. Jon Nehlsen, Chief Financial Officer

Melton Industries - Mr. Kenneth B Fairchild, Director, Government Affairs **NetIDEAS, Inc.** – Mr. Joseph Iannacone, Co-Founder and Vice President

New Jersey Institute of Technology (NJIT) - Dr. Donald H. Sebastian, Ph.D, Sr. Vice President for Research and Development

New Jersey Technical Procurement Assistance Center at NJIT (NJ PTAC) - Ms Dolcey E Chaplin, Esq., Statewide Director

Northrop Grumman – Ms. Gloria Pualani, Corporate Director, Socio-Economic Business Programs/Government Relations

Sea Box, Inc. – Mr. Robert A. Farber, Director of Contracts and Counsel

SMH International, LLC - Ms. Carol Hunt-Miller, Vice President for Business Development

Specialty Systems, Inc. - Mr. Emil Kaunitz, President & Chief Operating Officer **SRI Sarnoff** – Mr. Mark Clifton, Vice President, Products and Services Division / General Manager, Princeton

Temple University - Dr. Ken Blank, Senior Vice Provost for Research and Graduate Education

Trade Unions – Assemblyman Wayne DeAngelo, President of the Mercer and Burlington Counties Building and Construction Trades Council, Representative in the New Jersey General Assembly

Ve Source, LLC – Mr. Christopher R. Neary, Chief Executive Officer

APPENDIX F – SUMMARY OF INDUSTRY ROUNDTABLE AT SANTA CLARITA, CALIFORNIA

Overview

On January 8, 2012, Members of the Panel of Business Challenges in the Defense Industry traveled to Santa Clarita, CA to meet with members of the local defense industry. The Panel held a roundtable discussion on January 9, 2012 at the Santa Clarita City Hall with representatives from the local defense industry (see Appendix A for a list of participants).

Following the discussion with local defense industry, the delegation met with Air Force officials, and toured facilities at Air Force Plant 42, a GOCO facility managed by the Air Force. Plant 42 is the second-largest employer in the Antelope Valley, after Edwards Air Force Base, and consists of an Air Force operated airfield and 3.2 million square feet of contractor operated industrial facilities which are leased via separate agreements to Boeing, Lockheed-Martin and Northrop Grumman. The plant has a replacement value of $1.1 billion and is involved in manufacturing, maintenance, modification and testing of military aircraft such as the B-2, F-22, F-35, U-2 and Global Hawk.

Members in Attendance

HASC Chairman "Buck" McKeon
Panel Chairman Bill Shuster
Panel Ranking Member Rick Larsen
Ms. Colleen Hanabusa

Point of Discussion

- **Defense Contract Audit Agency** – several participants expressed frustration with DCAA's failure to close out incurred cost audits in a timely manner. One company was last audited in 2005 and the audit was still open, costing the company an estimated $3-4 million in lost business over the last six years. The participant noted that the contracting officers requested indirect rate audits but DCAA was non-responsive and the company was prohibited from moving forward from a successful SBIR Phase II contract because the audit was still open. It was suggested that the Panel should consider mandating maximum turn-around times for audits such as 60 days for rate audits, and 6 months for incurred cost audits. It was also suggested that contracting officers should be allowed to issue letter contracts so that they can proceed with a contracting action while an audit is still open and make adjustments, if necessary, after the audit is closed. Another participant felt that turnover and inexperience with DCAA auditors was part of the problem. It was stated that every year they get a new auditor and they have to start all over because the new auditor uses different processes and has different audit requirements. In order to address this issue, it was suggested that DCAA should be required to report performance metrics in order to highlight regional shortcomings and more uniform [military member] involvement at DCAA was needed to balance the inexperienced civilian workforce.
- **Technology Development and Transition** – a participant noted that the SBIR program was a wonderful program but there is no system to help a small business get to production. The participant posed the question "How do we get to the guys that want to buy our product when FFRDCs, universities, and the primes are all shooting at us?" It was also suggested that the Rapid Innovation Fund[231] (RIF) is not being successfully implemented and that more funding was needed

and the Panel was cautioned that "if you cut these budgets you're not going to see innovation" because the primes will suck up the [research and development] money and go on life-support. A participant went on to say that "with the elimination of earmarks, more innovative technologies will have a tougher time getting across the "valley of death". As a result, significantly more funding at the RIF [development] phase and more importantly funding focused on the "qualification" phase is required. This qualification phase falls between the two programs [science and technology and acquisition funding programs] really and can also cost more money than is typically available under either program."

- **International Trafficking of Arms Regulations)** – many of the participants expressed issue with ITAR and export licensing. It was stated that ITAR is "very detrimental" because other companies will not even consider you for [partnering or subcontracting] if you have to maintain ITAR compliance. It was also stated that European companies are not buying US products because of ITAR issues, but they can sell their products here. It was noted that while ITAR itself has not been changed recently, it really began being enforced in the last 3-4 years. It was stated that many of the big primes have been fined, as have individuals. It was suggested that in any reform being considered, ITAR should be applied by product line. An example was given that on a particular contract to provide DOD air conditioners, that the [basic] air conditioner was required to be ITAR compliant simply because DOD wanted it to be painted with chemically resistant paint. A participant also expressed frustration with the process and stated that there is no avenue for business to understand how to navigate the ITAR processes. Software was also raised as an area of major concern for ITAR certification. A participant suggested that the Panel should consider recommending that a company should be certified as "ITAR-compliant" rather than requiring certification by individual product lines.
- **Access to the Customer** – a participant stated that one of the major problems is getting to/through the program offices. A participant noted that small businesses are often not invited to participate in shows and conferences because program managers want to go with the incumbent or other "trusted agent" such as a large prime. It was acknowledge that small business fraud is still a problem. Another participant referenced an Army initiative to have a two-stage proposal

process. In this program a "pre-proposal" would be issued that gives industry insight in to what DOD wants, and allows industry to submit a 1-2 page response. At that point, DOD can determine which companies should be invited for full proposal. This approach was lauded as saving industry time and resources and as being good for small business because they can find out they are not suited for the procurement before wasting resources to complete a full-proposal.

Industry Roundtable Participants

AeroVironment - Tim Conver, Chairman & Chief Executive Officer
ADI, - John Cave, President
Advatech Pacific - Jay P. Ebersohl, P.E., President & Chief Executive Officer
Applied Companies - Joe Klinger, Regional Business Director
Arcata - Tim Wong, President
Circoil - Howard Lind, President and Chief Executive Officer
Crater Industries - Matt Donaldson, President
Delta Scientific - Harry Dickinson, Senior President
Electricore - Ilker "Ike" Bayraktar, President and Chief Executive Officer and Deborah Jelen
Exquadrum - Eric Schmidt, VP Engineering
Quallion - Paul Beach, President
Semtech - Charles Harper, Senior Vice President
Senior Systems Technology - Tim Morrissey, President/ Chief Executive Officer, Senior Systems
Triumph - Bill Boyd, President

Appendix G – Summary of Industry Roundtable at Honolulu, Hawaii

Overview

On January 9, 2012, Members of the Panel of Business Challenges in the Defense Industry traveled to Honolulu, HI to meet with members of the local defense industry. The Panel held a roundtable discussion on January 10, 2012 at the State Capitol building with representatives from the local defense industry (see Appendix A for a list of participants). Following the discussion with local defense industry, the delegation received briefings from the Commander, US Pacific Command, and from the Commanders of the Pacific Fleet, Marine Forces Pacific, and Pacific Air Forces. In addition, the delegation received a briefing on submarine operations in the Pacific and toured the USS Texas (SSN-775), a Virginia-class submarine. The delegation also toured Pearl Harbor Naval Shipyard and discussed the role of private industry in the shipyard operations.

Members in Attendance

Chairman Bill Shuster
Ranking Member Rick Larsen
Ms. Colleen Hanabusa

Points of Discussion

- **Small Business Innovative Research Program** – several participants commented on the SBIR program:
- A participant stated that the SBIR Phase III[232] needs to be restructured and that contracting officers do not know how to execute a Phase III. It was suggested that there should be a central point of contact or ombudsman for SBIR to help educate contracting officers. It was also suggested that a percentage of the defense budget, drawn from acquisition programs, should be set aside solely for Phase III.
- Another participant stated that after successful completion of Phase I[233], they waited 9 months for the Phase II[234]. It was remarked that program managers are not graded on moving from Phase I to Phase II and that small business are often told to just go find a large prime to transition technology. It was suggested that there should be a mechanism for measuring transition performance in the SBIR program.
- Another individual commented that program offices appear to be risk averse and there is no motivation to develop SBIR topics. It was also stated that only those companies physically located near the program office can influence SBIR topic development and unless you are already an established part of a bigger program there is no opportunity and no ability to transition.
- **Acquisition System/Processes** – several participants commented on the acquisition system and processes:
- One participant stated that small businesses are not equipped to deal with the bureaucracy of the DOD acquisition system. It was suggested

that there needs to be a middleman in order to connect small business to requirements developers.

- A participant referenced an Army memo regarding sole-source contracting and believes it is being misinterpreted by contracting officers and needs to be clarified or rescinded. [Note: the committee obtained a copy of this memo and is investigating the issue.]
- It was stated that a program manager remarked to a participant that they "don't do" sole source contracts and it would take at least 18 months to implement even a very small contract.
- It was also stated that DOD's preference for "low price/technically acceptable" solutions is not good for the Warfighter.
- One of the participants commented that contracting officers are often overruled by the lawyers. It was suggested that an ombudsman at the COCOM [Combatant Command] level was needed to address these issues.
- Another participant remarked that the government is becoming more risk averse than industry and an acquisition system driven by cost, schedule and performance drives low-risk approaches. The participant went on to say that the if the proposed budget cuts are put in place, core acquisitions and the government labs will preserve themselves and cut all external activities.
- A participant stated that some in the DOD have the view that "you're a small business --- take a small role."
- Another participant commented that large contracting vehicles like the Navy's Multi-ship Multi-Option contract for ship repair can enable large businesses to provide overhead and cost accounting for the small businesses on the contract.
- A participant applauded the work done by Lt Gen Thiessen, Commander, Marine Forces Pacific (MARFORPAC), for his efforts in reaching out to small businesses, specifically through the MarForPac Experimentation Center (MEC)[235]. It was stated that they [MARFORPAC] have a great model that is not currently replicated anywhere else.
- One participant remarked that it is very hard for a small business to find out about partnering opportunities with foreign companies and noted that, based on their location in the Pacific, there are opportunities for Hawaiian companies to partner with foreign firms.
- A participant recounted an experience negotiating with DOD on a commercial service and even after submitting their "best and final",

the contracting officer came back to them wanting a break out of each line item in order to negotiate the price. It was remarked that the negotiations take time and money, and are unnecessary for a procurement of a commercial service.

- Several participants articulated that testing and evaluating their technologies with the customer was critical for developing the right product on time and on budget. Part of the innovative technology "valley of death"[236] is due to lack of test and evaluation funding.
- **Disruptive Innovation** – one participant cited the failure of the Kodak Company and advocated for new capabilities in manufacturing. It was noted that manufacturing is very difficult, but companies need to look to the future and new methods need to be created and the government needs to make sure we retain this critical capability in various parts of the country. It was also stated that there is no incentive for disruptive innovations. Another participant remarked that innovation occurs one of two ways: 1)by accident, or 2) because of a champion with a passion. The system is not set up to incentivize innovation.
- **Technology Development and Transition** – a participant commented that DOD field activities, such as SPAWAR [Space and Naval Warfare Systems Command], accomplish work when it could be given to small businesses. It was alleged that some of these activities are competing with industry to sending letters to state agencies saying they are the only provider of a service.
- **International Trafficking of Arms Regulations** – several participants addressed various challenges experience with unnecessary ITAR restrictions and burdensome processes. They all commented that ITAR needs to be reformed.

Industry Roundtable Participants

Aina Kai Environmental (AKE) - Mun-Won Chang, President & Chief Executive Officer

BAE Systems - Alan Hayashi, Support Solutions Director of Public Relations and Policy Advocacy

DreamHammer - Larry Osborn, Executive Vice President and Chief Strategy Officer and Member of the Board of Directors

General Atomics - Dr. John Parmentola, Senior Vice President for Energy and Electromagnetic Systems

Hawaiya Technologies, Inc. (HTI) - Paul Schultz, President and Chief Executive Officer

Ke'aki Technologies - Philip Kahue

Matson Navigation Company - Vic Ancogo, Senior Vice President-Pacific

Oceanit - Pat Sullivan, President and CEO and Jan Sullivan, Chief Operating Officer
Referentia Systems Incorporated - Nelson Kanemoto, Founder, President & Chief Executive Officer
University of Hawaii at Manoa - Dr. Jim Gaines, Vice President for Research

APPENDIX H – SUMMARY OF INDUSTRY ROUNDTABLE AT SAN DIEGO, CALIFORNIA

Overview

On January 10, 2011, Members of the Panel of Business Challenges in the Defense Industry traveled to San Diego, CA to meet with members of the local defense industry. The Panel held a roundtable discussion at the Admiral Kidd Club, Naval Base Point Loma, CA, on January 11, 2012 with representatives from the local defense industry (see Appendix A for a list of participants). In additional to the roundtable meeting with local defense industry, the delegation toured General Dynamics Marine Systems' NASSCO shipyard and boarded USNS Medgar Evers (T-AKE 13), a Lewis and Clark-class dry cargo ship that is nearing completion. The delegation also boarded USS Freedom (LCS-1), the lead ship in the Freedom-class of littoral combat ships and also toured the trainer facilities and simulators used to prepared crews for LCS operations. The delegation concluded the visit with a tour and briefings of Space and Naval Warfare Systems Command (SPAWAR). As one of three Department of Navy major acquisition commands, SPAWAR focuses on information dominance and is engaged in acquiring, installing, delivering and maintaining advanced information technology capabilities to the fleet, regardless of platform, to keep Warfighters one step ahead of adversaries.

Members in Attendance

Chairman Bill Shuster
Ranking Member Rick Larsen
Ms. Colleen Hanabusa
Mr. Jon Runyan
Mr. Duncan Hunter
Ms. Susan Davis

Points of Discussion

- **San Diego Advanced Defense Technology Cluster** - a participant commented that the Cluster, funded through the SBA, works to aid small business in transitioning technology and helps to grow small businesses. In its second year, it is working with 30 companies in the San Diego area. It was remarked that it is a very good program to support SBIR transition.
- **Finances** – a participant noted that the 15% holdback [retention on 15% of contract value for a stipulated time to ensure contract completion and payment of all parties] on profit is really hard on small business.
- **Defense Contract Audit Agency** - a participant commented that DCAA is underfunded and is very slow to close out a contract. It was also stated that waiting nine months on DCAA [to complete an audit] can put a small business out of business.
- **Small Business Innovative Research Program** - one participant remarked that there needs to be a structured and resourced Phase III program to provide companies resources to field technologies. Participants cautioned the Panel about the inclusion of venture capitalists in the SBIR program and expressed concern that venture capitalists could take technology developed through the SBIR program and sell to foreign entities. [Note: The Fiscal Year 2012 NDAA re-authorized the SBIR program for 6 years and authorized the participation of venture capitalists in the program.]
- **Congressionally-directed Funding** - a participant commented that Congress has a responsibility and is abdicating it by not allowing earmarks. It was stated that losing it [ability to direct funding] has a huge negative effect on small businesses. The participant stated that without earmarks, DOD does not know what [technologies] are out there. Another participant remarked that there is a need for Congressionally-directed funding, particularly in this tough economic time.
- **Technology Transition** – one participant remarked that it is not just a need for money for development, but also to support getting a technology fielded. It was suggested that start-up money for these areas in the high-technology sector can assist a lot of laid-off people. Another participant stated that the primes are still being allowed to act

as systems integrators when small business could be involved in manufacturing given the opportunity.

- **Contracting Vehicles** - a participant expressed frustration about a program they had met requirements and objectives on, but the program office as unable to obtain a contracting vehicle. It was stated that you cannot become a program of record until successful completion of an operational evaluation and you can't do that if they can't provide a contracting vehicle.

  Another participant noted that their company would not subcontract, but noted that typical procurement officers want to use existing contracting vehicles to "pass through" work to the small business. It was remarked that this was being done because it simply takes too long to go through the normal acquisition process. Another participant commented that on one particular program the government asked a company to purchase $10 million in radios via an existing contract [for other work] because it would take 9 months to get a contracting vehicle in place to procure the radios and the capability was needed in 6 months.
- **Intellectual Property** – a participant stated that the large primes don’t want small businesses to innovate and another participant stated that anyone that wants to do business with you wants your technology. Both agreed that more needs to be done to protect the intellectual property of small businesses.
- **Access to the Customer** - a participant commented that small businesses are isolated from manufacturers and they can't talk directly with the person with the problem.

  Another participant suggested that small business needs a structure that allows business to go directly to DOD. A third participant noted that he had to get direct assistance from Congressman Hunter to get invited to participant in the acquisition process. This participant described the acquisition process as being ineffectual and not sympathetic to small businesses.
- **International Trafficking of Arms Regulations** - one participant stated that it seemed individual opinions often drive licensing decisions, making it an inequitable system, and that the export controls are very complex and hard to navigate. Another stated that there were no major issues with ITAR, while a third commented that companies are restricted from selling their technologies so foreign partners are just buying the technology elsewhere. Another participant

remarked that the time to get through the process is very long and it compromises the ability to get the mission done. A participant also commented that ITAR negatively affects universities [doing research for the government] as well and remarked "we're tying ourselves in knots for nothing."

- **DOD Culture** - a participant commented that innovative or creative program offices can "get slapped upside the head by leadership". It was implied that more flexibility needs to be given to program offices in order to deal with many of the issues in the acquisition process.

Industry Roundtable Participants

Allylix - Seth Goldblum, Vice President of Business Development

Aurora Aerospace - Myles Newlove, President & Chief Executive Officer

Cubic Corporation - Mike Kelly, Vice President, Strategy & Development, Cubic Defense Applications, Inc.

Digibeam Corporation - Mike Zani, Chief Executive Officer

East County Economic Development Council - Jo Marie Diamond, President & Chief Executive Officer

ES3 - Teri Sgammato, Chief Executive Officer

GET Engineering - Greg MacNeil, Chief Executive Officer

JCI Metal - Marcel Becker, Chief Executive Officer

NSM Surveillance - Andy Berdy, President

Port of San Diego Ship Repair Association - Derry Pence, Chief Executive Officer

Sapphire Energy - Tim Zink/Denise Gitsham, Vice President of Corporate Affairs/Director of Corporate Affairs and Legislative Counsel

San Diego Advanced Defense Technology Cluster (SDADT) - Lou Kelly, Program Director – San Diego State University

San Diego Composites - Rob Kolozs, President

San Diego State Research Foundation - W. Tim Hushen, Associate Executive Director, Research Advancement

San Diego Supplier Development Council - Paul Hollenbach, Chairman, Corporate Development Committee

Scripps Institution of Oceanography, University of California-San Diego - Bruce Applegate and Kathleen Ritzman

Smart Electronics and Assembly - Yvonne Johnson

University of California, Irvine - George Peavey, Director of the Medical Free Electron Laser Program

University of California, San Diego - Byron Washom, Director of Strategic Energy Initiatives

Appendix I – Summary of Industry Roundtable at Jupiter, Florida

Overview

On February 21, 2012, Members of the Panel of Business Challenges in the Defense Industry traveled to Jupiter, Florida to meet with members of the local defense industry.

The Panel held a roundtable discussion on February 21, 2012 at the Florida Atlantic University's MacArthur Campus with representatives from the local defense industry (see Appendix A for a list of participants). Following the discussion with local defense industry, the delegation met with the Northern Palm Beach County/Treasure Coast Chamber of Commerce to hear the concerns and challenges that small business are facing in Northern Palm Beach County and the Treasure Coast.

Members in Attendance

Chairman Bill Shuster
Mr. Allen West
Small Business Committee Chairman Sam Graves

Point of Discussion

- **Competition**– a participant suggested that more needs to be done to force DOD to use competitive acquisition strategies, particularly in the sustainment of weapon systems. It was stated that many program managers prefer to issue a sole-source contract to the original equipment manufacturer when there are other companies that could compete for the work. It was suggested that increased competition would reduce program costs rather than DOD having to buy fewer of items it needs to stay under the budget. Another participant felt that program offices failed to do sufficient market research and tended toward "sole-source for life" approaches to procurement.
- **Bundling of Contracts** – a participant stated that the military services are bundling or consolidating a lot of their contracts and it eliminates

competition and makes it impossible for small businesses to perform the large scope of work required in many bundled contracts. It was suggested that DOD could drive more competition into these bundled contracts by using blanket purchase agreements.

- **International Trafficking of Arms Regulations**– Several participants indicated that ITAR was negatively impacting the US industrial base. One participate stated that ITAR is selectively applied while another indicated that small businesses "run away" from defense contracts because of a fear that something they produced will mistakenly end up in the wrong hands and they will be held accountable.
- **Consistency in Workloading the Industrial Base** – several participants suggested that DOD needs to work with industry to provide a level workload.
  - One participant said that DOD would issue short average production contracts (3-4 months) for their product and it would drive violent swings in workload and make it difficult, especially for small business, to manage their work force. It was acknowledged that DLA's recent involvement has improved the situation to some degree but more needs to be done to provide industry with a consistent demand signal.
  - It was stated that laying off and then rehiring workers drives up costs due to training, scrap/rework and a less reliable workforce.
  - Another participant stressed that industry needs long-term production commitments so that they can maintain a viable, reliable supply base. It was stated that having to restart production and recertify processes was inefficient and costly.
  - A producer of a major weapon system stated that multi-year contracts were a very good way to keep stability in the supply base but that there was a recent trend toward single year contracts on programs that had previously been executed under multi-year contracts; concerns were raised over budget reductions and funding instability.
- **Berry Amendment** – support for the Berry Amendment was expressed.
- Defense Logistics Agency -
  - A participant cited problems with DLA and stated that the transition from "the program office to DLA is a trainwreck." The participant commented that bureaucrats in DLA were not

subjected to performance evaluations, had no technological skills and would not be successful if they were in the private sector. It was recommended that more needs to be done to drive collaboration between the government and industry.

- Another participant said that DLA is not responsive to industry's suggestions and that they are territorial and like having authority.

- Defense Contract Management Agency and Defense Contract Audit Agency –
    - A participant who was working to become a prime contractor commented that DCAA and DCMA were working to see which of them could make it harder on industry. It was stated that they have too many branches and that none of them talk to any of the others.
    - Another participant noted that DCAA had become more adversarial over the last two years and they are becoming increasingly inefficient. Frustration was expressed over their audit methodologies even when dealing with a known, proven product and a known supplier.
    - A participant commented that DCMA should not be involved in commercial item procurements.
- **Consolidation of Contracting Offices** – a participated noted a trend toward regional consolidation of contracting offices and stated that purchase decisions that can and should be made locally are now being made regionally and this is significantly increasing the bureaucracy.
- **Contract Methodology** – a participant recommend DOD needs to take a "best value" approach rather than a "low price" approach when buying things for the Warfighter.
- **Security Clearances** – a participant cited difficulties in retaining (transferring) a security clearance once he left a large defense firm to start his own business. It was stated that DOD required a contract to validate the need for the clearance, but the solicitation for the contract required offerors to have a clearance.
- **Access to Facilities/Equipment** – it was suggested that small businesses be provided a "right of use" in order to enable them to gain access to government-owned equipment and tooling that had already been procured for other DOD programs during the development and testing phases.
- **Acquisition Workforce/Culture** – a participant stated that the acquisition workforce and contracting officers are not consistent in

their actions, training and capabilities. It was also noted that the DOD's acquisition workforce is very risk averse. Another participant stated that industry needed a direct line to the customer. It was also stated that the rules and regulations are extremely complex and are always changing.

- **Counterfeit Parts** – a participant commented that it is very difficult to compete when they are being undercut by counterfeits. It was indicated that DOD's preference for low price vs. best value is driving an increase counterfeits making it into the DOD supply chain.
- **Small Business Innovative Research Program** – a participant asked that more funding be provided through the SBIR program and another suggested something be done to reduce the time between SBIR Phase I and SBIR Phase II.

Industry Roundtable Participants

Agilis - Frank O'Neill, President and Chief Executive Officer

All American Warrior Networks - Randall Spillers, Tim Natole and Tim Evard

ATK Missile Products - Steve Beckel, Director of Advanced Propulsion Business Panel and Elizabeth Byrne, Senior Program Manager

Berla Corporation - Mike May, President and Chief Executive Officer and Katrin Ratassepp, President NCMA, South Florida Chapter

BRS Aerospace - Larry E. Williams, President and Chief Executive Officer **Cartwright & Associates** - Russell Cartwright, President and Chief Executive Officer

Chromalloy - David G. Albert, Vice President of External Affairs and Dr. Michael Beffel, Vice President of Operations

Cross Match -Thomas Buss, Senior Vice President of Strategic Initiatives and Kathy Hutton, Senior Vice President

Cyclone Power Technologies Incorporated - Doug Hutchinson, Vice President of Business Development

Dayton-Granger Incorporated - Gibbons D. Cline, President

Florida Atlantic University - Dr. Pierre-Philippe Beaujean, Associate Professor

Florida Turbine Technologies - Shirley Brostmeyer, Chief Executive Officer and Dr. Susan Cunningham, Director of Business Development and Marketing

Lockheed Martin - Richard Holmberg, Vice President

Masse Reaction - Arthur Masse, CEO and Nick Nicklas, Senior Vice President of Marketing

Nova Southeastern Oceanographic Center - Dr. Richard Dodge, Dean

Nova Southeastern University - Gary Margules, Vice President of Research and Technology Transfer

Sikorsky - John Fischetti, Director, Flight Demonstration Center for Sikorsky

Pratt Whitney Rocketdyne - Jim Maus, Director of Enterprise and General Manager

Source One Distributors - Mark Lano, President and Chief Executive Officer and Randy Webber, Chief Operations Officer

United Electronics Corporation - Shane Scanion, Chief Executive Officer and David York, President
Vutec Corporation - Howard Sinkoff, Chairman and Paul Passolacul

APPENDIX J – SUMMARY OF INDUSTRY ROUNDTABLE AT WASHINGTON, DC

Overview

On February 27, 2012, Members of the Panel of Business Challenges in the Defense Industry held a roundtable in Washington, D.C. with members of the local defense industry. The Panel met in Rayburn House Office Building, Room 2212 with representatives from the local defense industry (see Appendix A for a list of participants).

Members in Attendance

Chairman Bill Shuster
Ranking Member Rick Larsen
Mr. Jon Runyan
Ms. Betty Sutton
Mr. Allen West
Ms. Colleen Hanabusa

Points of Discussion

- **Defense Contract Management Agency** – a participant cited grave challenges with DCAA and indicated that his business was at risk due to actions of a prior employee and DCAA audit response.
- **Defense Logistics Agency** – several participants commented on the lack of accountability by DLA employees in terms of providing information to the contractor. It was also noted that DLA 'reverse auctions' did nothing to improve the quality and performance of the contractor. It was also stated that DLA's demand-based approach does little to provide industry visibility on requirements.

- **Inconsistency in Contracting across DOD** – a participant noted that each of the military services do things very differently even when procuring the same item. An example was cited that in buying the same product the Army used a FAR Part 12 (commercial item) contract and the Air Force used a FAR Part 15 (negotiated acquisition) contract; one service considered the procurement to be a services contract and the other service looked at it as an end-item procurement. It was suggested that DOD needs to development an enterprise wide approach to contracting to improve efficiency.
- **Supplemental Budget Items** – a participant noted that some contracts have been funding for over a decade from supplemental funding. They referenced an 'era of low price' versus value and commented on the lack of visibility and vision for the future when certain contracts do not receive budget lines.
- **Incentives** – several participants commented on the need for more incentives to receive a better product and have a more efficient acquisition and procurement process in place. They noted that incentives for both the contractors and for the government program managers would eliminate some of the risk aversion within the Department of Defense in awarding contracts to small and midsize businesses who have never worked with DOD before.
- **GSA Schedules** – a participant noted that the GSA schedule is a long and arduous process for a new business. To get on a GSA schedule, you have to have been in business for a minimum of two years.
- **Small Business Contracting Plans** – it was stated that primes are not held accountable for compliance with their Small Business Contracting plans and will often swap out subcontractors after winning the contract. It was suggested that primes should also take "per capita" cuts when reductions are necessary rather than being allowed to take profit for their shareholders while pushing the funding reduction down onto the subcontractors. A large business representative stated that primes needed to be able to swap out sub for a variety of reasons, to include performance. It was also noted that small business contracting goals are extremely complex and very difficult for primes and the government to track and monitor.
- **Access to the Customer** - a participant commented that small businesses are isolated from manufacturers and they can't talk directly with the person with the problem. Another participant suggested that small business needs a structure that allows business to go directly to

DOD. A third participant noted that he had to get direct assistance from Congressman Hunter to get invited to participant in the acquisition process. This participant described the acquisition process as being ineffectual and not sympathetic to small businesses.

- **Communication with Industry** - a participant remarked that "Industry Days" and the public "Questions and Answers" during the solicitation are completely ineffective because no one will speak up for fear of losing their competitive advantage or telegraphing what their approach may be. It was suggested that DOD needed to hold closed meetings with each contractor to provide for a better exchange of information. The GSA FedSim program was suggested as a model and it was stated that it was a "due diligence session" with contractors who had been awarded IDIQ contracts.
- **DOD Technology** – it was noted that DOD has not advanced its business systems and approaches as industry has. It was recommended that DOD consider moving toward a web-based procurement for commercial items much like the GSA website that has been created for government agencies to procure IT applications based on an Amazon.com approach (www.apps.gov).
- **Security Clearances** – a participant noted difficulties getting a facility clearance and suggested that the Defense Security Service needs to streamline Facility Security Office (FSO) training and take a different, more simple approach in training small businesses FSOs. It was also suggested that DOD should consider moving to a DOD-wide Common Access Card (CAC) because if a contractor has an Army contract and an Air Force contract they are only allowed one CAC card and the cards, which are critical to facility access and logging in to IT systems, will not work in both systems. A participant also suggested that small businesses should be provided access to other cleared DOD or prime contractor facilities rather than having to develop a cleared facility on their own.
- **Low-cost/Technically Acceptable vs. Best Value** – a participant indicated that DOD's preference for low-cost/technically acceptable contracting approaches provides no incentive for innovation. Another participated noted that when DOD moved from best value to low-cost. A third participant stated that there are no skills in the government to do best value assessments.
- **Timely Payment to Subcontractors** – a participant suggested that Program Managers need to be aware of and monitor payments to

subcontractors. An example was cited where an invoice for one company was bundled with 12 other companies and sent for payment. The entire invoice was held for payment because of performance of one of the 13 companies.

- **Prime – Sub Relationship** – a participant remarked that prime contractors will only flow-down information that they have to and they prefer to keep sub contractors out of the loop. Another noted that partnering with large companies is not easy because the prime will scope the workload on only pass down what they do not want to do --- a sub is at their mercy. Another participant remarked that the government uses primes as a "shield".
- **Tailored Logistics Contracts** – a participant noted that tailored logistics contracts used to support the war efforts are working well and concern was expressed that this type of contracting may go away as funding and resources are reduced.

Industry Roundtable Participants

Ability Worldwide Inc. – Suzanne Schwitalla, President and Chief Executive Officer
Advanced Manufacturing Technology Ventures – Brittany McCracken, General Manager
Aeronautical Systems, Inc. - Joe Gass, Director, Business Development
All Quality Spares – Joseph Garofano, President
Avue Technologies - Linda Rix and James Miller, Chief Executive Officers
Blue Storm Technology - Mary Lockhart, President & Chief Executive Officer
Ceradyne Armour Systems, Inc – Mark King, President
CyOptics - Stefan Rochus, Vice President, Marketing and Business Development
Earl Energy - Josh Prueher, Chief Executive Officer
Honeywell – Mike Marinshaw
Hydronalix, Inc. - Robert Lautrup
Progeny Systems Corporation – Walter Kitonis
TCOM – Richard Martin, Chief Executive Officer
Tensley Consulting, Inc. – Michael K. Hantke, Chief Executive Officer
Unisys – Brian Clark, Senior Unisys Business Development Executive
Yakabod - Scott Ryser, Founder & Chief Executive Officer

Appendix K – Department of Defense Comment

The Department of Defense was provided a draft of this report and was offered an opportunity to provide comment. Many of the DOD's comments help to clarify and strengthen the content of the report and, as such were incorporated into the final report. However, some of the comments received were specifically provided response to remarks that were recorded at the industry roundtables or in response to findings by the Panel. As such, the Panel felt it appropriate to include those responses in this Appendix.

- Agency Comment (in general) - The report contains a summary the roundtable discussions held at different locations throughout the United States. These summaries contain several references to DCAA. Many of the references to DCAA in the roundtables were issues that reflected negatively on the Agency (i.e., audits taking too long to complete, backlog of incurred cost audits, issues with effective communications). If the Panel thought it would be productive, the Agency would like to follow-up of the issues described in the report to ensure these concerns have been resolved.
- In response to the Panel finding that the Defense Logistics Agency (DLA) audit organization failed its peer review and some observers are concerned that defense audit organizations will also fail peer reviews, the Department responded with the following:

> The DLA OIG [Office of the Inspector General] has taken corrective actions to ensure compliance with Government Auditing Standards to include issuing revised policies and procedures for Internal Audit, Audit Follow-up, External Audit and our Quality Controls Audit Policy (QCAP). We have initiated a quality assurance function, provided training to all auditors and implemented a comprehensive human capital strategy to improve competencies. A recent Quality Assurance Review (QAR) assessed compliance with GAGAS requirement on auditors' professional competence through continuing professional education (CPE) for the two-year measurement period of FY 10 -FY 11. Overall, DLA OIG Audit Division has a system of quality control embedded in its policies and procedures over the GAGAS CPE requirements; and all 38 DLA OIG auditors, as of September 30, 2011, met the minimum CPE hour requirements for the two-year measurement period. Our reorganization is designed to ensure independence and separated the OIG function from Internal Review. Additionally, DLA has developed a more comprehensive stewardship process to ensure better identification of Enterprise Risks. The DODIG has already provided positive feedback on our reorganization, revised policies and training. We are confident that the corrective actions we have taken to address the previously identified deficiencies will be sufficient to pass our peer review later this year.

- In response to comments heard by the Panel in Mount Laurel, NJ on December 9, 2011, regarding a notice of solicitation cancellations, the Department responded that all cancellations "are posted on www. FEDBIZOPS.gov with solicitation number and point of contact. They also noted that "DLA takes very seriously all DOD small business goals, to include Service Disabled Veteran Owned companies, and aggressively works to provide small businesses contracting opportunities. DLA is working to enhance our Service Disabled Veteran Owned Small Business (SDVOSB) program with the following initiatives:
- Current National Stock Numbers (NSNs) reserved for SDVOSB set-asides is 7,027.
- Provide additional SDVOSB training to DLA Acquisition Workforce.
- Work with various organizations to identify new capable SDVOSB Suppliers (SDVOSB Council, outreach events such as National Veteran Small Business Conference and Expo and Veteran Entrepreneur Training Symposium).
- Identify additional DLA Enterprise-wide contracting opportunities in areas where the majority of SDVOSB firms are, such as program management, consulting and information technology. For FY12 we are on track with a credible plan to meet this goal."
- In response to comments heard by the Panel in Jupiter, FL, on February 21, 2012, regarding short-term contracting, the Department responded that "unfortunately, due to emergent mission requirements, DLA has had to issue short term bridge contracts to meet customer needs. We have an aggressive acquisition planning program in place to put long term contracts in place which should allow for a more consistent demand signal. This should allow contractors to better manage workload and their respective workforce."
- In response to comments heard by the Panel in Jupiter, FL, on February 21, 2012, the Department responded that "DLA is committed to more frequent and consistent industry engagement. A priority for DLA in FY12 is effective collaboration between the Government and industry with a goal to increase productivity and performance for the Warfighter."
- In response to comments heard by the Panel in Washington, DC, on February 27, 2012, regarding reverse auctions, the Department responded that "reverse auctions are used to promote aggressive materiel savings for items that are normally commercial and procured

on a fixed price basis. Quality and performance trade-offs are normally used in best-value procurement."

APPENDIX L – ACRONYM LIST

CCL	Commerce Control List
CRADA	Cooperative Research and Development Agreement
CSBA	Center for Strategic and Budgetary Assessments
CSIS	Center for Strategic and International Studies
DARPA	Defense Advanced Research Projects Agency
DAWDF	Defense Acquisition Workforce Development Fund
DBB	Defense Business Board
DCAA	Defense Contract Audit Agency
DCMA	Defense Contract Management Agency
DDTC	Directorate of Defense Trade Controls
DFARS	Defense Federal Acquisition Supplement
DHS	Department of Homeland Security
DIB	Defense Industrial Base
DLA	Defense Logistics Agency
DMAG	Deputy's Management Action Group
DOC	Department of Commerce
DOD	Department of Defense
DOE	Department of Energy
DPAS	Defense Priorities and Allocations System
EAA	Export Administration Act
FAR	Federal Acquisition Regulation
FFRDC	Federally Funded Research and Development Center
FPDS	Federal Procurement Data System
FTTA	Federal Technology Transfer Act
FY	Fiscal Year
GAO	Government Accountability Office
GOCO	Government-Owned, Contractor-Operated
GOGO	Government-Owned, Government-Operated
GSA	General Services Administration
HASC	House Armed Services Committee
HUBZone	Historically Underutilized Business Zone
IBIF	Industrial Base Innovation Fund
IG	Inspector General

IR&D	Independent Research and Development
IT	Information Technology
ITAB	Information Technology Acquisition Board
ITAR	International Trafficking of Arms Regulation
JB MDL	Joint Base McGuire-Dix-Lakehurst
ManTech	Manufacturing Technology
MIBP	Manufacturing and Industrial Base Policy
NDAA	National Defense Authorization Act
NRC	National Research Council
OEM	Original Equipment Manufacturer
OSBP	Office of Small Business Programs
OSDBU	Offices of Small and Disadvantaged Business Utilization
PAIR	Priority Allocation of Industrial Resources
PCR	Procurement Center Representative
PTAC	Procurement Technical Assistance Center
R&D	Research and Development
RDT&E	Research, Development, Test and Evaluation
RIF	Rapid Innovation Fund
S&T	Science and Technology
S2T2	Sector-by-Sector and Tier-by-Tier
SBA	Small Business Administration
SBIC	Small Business Investment Company
SBIR	Small Business Innovative Research
SDVOSB	Service Disabled Veteran Owned Small Business
SDB	Small Disadvantaged Business
SPAWAR	Space and Naval Warfare Systems Command
STEM	Science, Technology, Engineering and Math
UARC	University Affiliated Research Center
USD(AT&L)	Under Secretary of Defense for Acquisition Technology and Logistics
USML	United States Munitions List

End Notes

[1] U.S. Government Accountability Office, Warfighter Support: DOD's Urgent Needs Processes Need a More Comprehensive Approach and Evaluation for Potential Consolidation, March 2011.

[2] David Walker, "America's Imprudent and Unsustainable Fiscal Path," Defense Acquisition Technology and Logistics, April 2006.

[3] The Department of Homeland Security defines the Defense Industrial base as the "DOD, government, and the private sector worldwide industrial complex with the capabilities of performing research and development, design, production, delivery, and maintenance of military weapons systems, subsystems, components, or parts to meet military requirements." See Department of Homeland Security: Defense Industrial Base Sector: Critical Infrastructure and Key Resources. Found at: http://www.dhs.gov/files/programs /gc_1189165508550.shtm.

[4] Ibid.

[5] Barry Watts and Todd Harrison, Center for Strategic and Budgetary Assessments, "Sustaining Critical Sectors of the U.S. Defense Industrial Base", 2011, p. xii.

[6] Ibid, p.53.

[7] Barry Watts in testimony before the Panel on October 24, 2011.

[8] Fred Downey in oral testimony before the Panel on October 24, 2011.

[9] Brett Lambert in oral testimony before the Panel on November 1, 2011.

[10] This report can be found at: http://dbb.defense.gov/pdf/Industrial_Base_ Task_ Group_Report.pdf. In addition, the DBB has conducted a series of studies related to this, which can be found at: http://dbb.defense.gov/reports.shtml.

[11] Pierre Chao in testimony before the Panel on October 24, 2011.

[12] For more on the SBIR program, see generally Congressional Research Service Report R96-402, Small Business Innovation Research (SBIR) Program, by Wendy H. Schacht.

[13] Congressional Research Service Report R40764, Department of Defense Contractors in Iraq and Afghanistan: Background and Analysis Moshe Schwartz, September 21, 2009.

[14] Brett Lambert in oral testimony before the Panel on November 1, 2011.

[15] This report can be found at: http://www.acq.osd.mil/mibp/docs/annual_ind_ cap_ rpt_to_congress-2011.pdf.

[16] Section 2504, Title 10, United States Code

[17] In written testimony provided to the Panel on November 1, 2011, p. 8

[18] Ibid, p. 8.

[19] See http://www.whitehouse.gov/the-press-office/2012/03/16/executive-order-national-defense-resources-preparedness.

[20] 15 U.S.C. § 631.

[21] 15 U.S.C. § 644(a).

[22] U.S. Government Accountability Office, Small Business Administration: 7(a) Loan Program Needs Additional Performance Measures, GAO-08-226T, November 1, 2007, pp. 3, 9-11; and Veronique de Rugy, Why the Small Business Administration's Loan Programs Should Be Abolished, Washington, DC: American Enterprise Institute for Public Policy Research, AEI Working Paper #126, April 13, 2006, http://www.aei.org/docLib/2006 0414_ wp126. pdf.

[23] Veronique de Rugy, Why the Small Business Administration's Loan Programs Should Be Abolished, Washington, DC: American Enterprise Institute for Public Policy Research, AEI Working Paper #126, April 13, 2006, http://www.aei.org/docLib/20060414_wp126.pdf.

[24] Candida Brush and Robert D. Hisrich, "Women-Owned Businesses: Why Do They Matter?," in Are Small Firms Important? Their Role and Impact (Boston: Kluwer Academic Publishers, 1999), pp. 111-127; and John Sibley Butler and Patricia Gene Greene, "Don't Call Me Small: The Contribution of Ethnic Enterprises to the Economic and Social Well-

Being of America," in Are Small Firms Important? Their Role and Impact (Boston: Kluwer Academic Publishers, 1999), pp. 129-145.

[25] Robert W. Fairlie and Alicia M. Robb, "Disparities in Capital Access between Minority and Non-Minority-Owned Businesses: The Troubling Reality of Capital Limitations Faced by MBEs," Washington, DC: U.S. Department of Commerce, Minority Business Development Agency, January 2010, pp. 3-5, 8, 17-23, http://www.mbda.gov/sites/default/files /DisparitiesinCapitalAccessReport.pdf.

[26] U.S. Small Business Administration, "FY2012 Congressional Budget Justification and FY2010 Annual Performance Report," Washington, DC, 2011, p. 1, http://www.sba.gov/sites /default/files/FINAL%20FY%202012%20CBJ%20FY%202010%20APR.pdf. Also, see U.S. Small Business Administration, "The Small Business Economy, 2010: A Report to the President," Washington, DC, 2010, pp. 26, 27, http://www.sba.gov/sites/default /files/sb_econ2010.pdf.

[27] U.S. Small Business Administration, Office of Advocacy, Small Business Economic Indicators for 2003, Washington, DC, August 2004, p. 3; and Brian Headd, "Small Businesses Most Likely to Lead Economic Recovery," The Small Business Advocate, vol. 28, no. 6 (July 2009), pp. 1, 2.

[28] 15 U.S.C. § 632(a).

[29] 15 U.S.C. § 632(a)(3).

[30] Pursuant to the Small Business Jobs Act of 2010, P.L. 111–240 (Sept 27, 2010), SBA must conduct a detailed review of all size standards and to make appropriate adjustments to reflect market conditions, and at least one-third of all size standards must be reviewed every 18 months.

[31] 74 Fed. Reg. 53940 (Oct. 21, 2009).

[32] 77 Fed. Reg. 7490 (Feb. 10, 2012).

[33] 13 CFR §107.700; 13 CFR §107.710; 13 CFR §301(c)(2); and 13 CFR §301(c)(1).

[34] According to the SBA's Office of Advocacy, small businesses "represent 99.7 percent of all employer firms[; e]mploy half of all private sector employees[; p]ay 44 percent of total U.S. private payroll[; g]enerated 65 percent of net new jobs over the past 17 years[; and c]reate more than half of the nonfarm private GDP." http://web.sba.gov/faqs/faqIndex All.cfm?areaid=24.

[35] U.S. Small Business Administration, Department of Defense Procurement Scorecard FY 2010 available at http://www.sba.gov/sites/default/files/files/FY10%20SB%20Procurement% 20Scorecard_FINAL_DOD.pdf.

[36] DOD spent $375.5 billion out of a federal total of $536 billion. See http://usaspending.gov/ .

[37] 15 U.S.C. § 644(a).

[38] 15 U.S.C. § 637(d). H.R. REP. NO. 95-949, at 5 (1978).

[39] See, e.g., 13 C.F.R. § 125.2.

[40] See, e.g., 15 U.S.C. § 644(k)(1)-(10).

[41] U.S. Government Accountability Office, Improvements Needed to Help Ensure Reliability of SBA's Performance Data on Procurement Center Representatives (GAO-11-549R) (2011); Government Accountability Office, Small Business Contracting: Action Needed by Those Agencies Whose Advocates Do Not Report to Agency Heads as Required, (GAO-11-418) (2011).

[42] U.S. Small Business Administration, "The Small Business Economy, 2010: A Report to the President," Washington, DC, 2010, pp. 2, 5, 21, 22, http://www.sba.gov/sites/default/files /sb_econ2010.pdf.

[43] U.S. Small Business Administration, Office of Advocacy, "Small Business Quarterly Bulletin: Second Quarter 2011," Washington, DC, http://www.sba.gov/sites/default/files/files /SBQB_2011q2.pdf.

[44] Ibid.

[45] U.S. Small Business Administration, Office of Advocacy, "Frequently Asked Questions about Small Business Finance," Washington, DC, September 2011, p. 4, http://www.sba.gov/sites /default/files/files/Finance%20FAQ%208-25- 11%20FINAL%20for%20web.pdf; and Federal Reserve Board, "Senior Loan Officer Opinion Survey on Bank Lending Practices," Washington, DC, http://www.federalreserve.gov /boarddocs/SnLoanSurvey/.

[46] U.S. Small Business Administration, "Small Business Procurement Scorecards," Washington, DC, http://www.sba.gov/content/small-business-procurement-goaling-scorecards.

[47] U.S. Congress, Senate Committee on Banking, Housing, and Urban Affairs, Small Business Legislation - 1974, hearing on S. 3137 and S. 3138, 93rd Cong., 2nd sess., March 13, 1974 (Washington: GPO, 1974), p. 19.

[48] U.S. Small Business Administration, "About Office of Surety Bond Guarantees," Washington, DC, http://www.sba.gov/aboutoffices-content/1/2891/about-us. Ancillary bonds are also eligible if they are incidental and essential to a contract for which SBA has guaranteed a final bond. A reclamation bond is eligible if it is issued to reclaim an abandoned mine site, and for a project undertaken for a specific period of time.

[49] U.S. Small Business Administration, "Surety Bonds," Washington, DC, http://www.sba.gov /category/navigationstructure/loans-grants/bonds/surety-bonds.

[50] Ibid.

[51] U.S. Small Business Administration, "Surety Bonds: Explained," Washington, DC, http://www.sba.gov/content/surety-bondsexplained.

[52] Performance bonds may be less than 100% provided that the contracting officer determines that a smaller amount will adequately protect the government. 40 U.S.C. § 3133(b)(2).

[53] U.S. Small Business Administration, "Standard Operating Procedure: Surety Bond Guarantee Program," SOP 50 45 2, Washington, DC, effective March 8, 1999, p. 7, http://www.sba.gov/sites/default/files/sop5045.pdf.

[54] U.S. Congress, House Banking and Currency, Housing and Urban Development Legislation - 1970, 91st Cong., 2nd sess., June 5, 1970 (Washington: GPO, 1970), p. 351.

[55] Ibid.

[56] U.S. Congress, House Committee on Small Business, Subcommittee on SBA Oversight and Minority Experience, Selected Small Business Administration Programs and Activities, 94th Cong., 2nd sess., February 24, 1976, H. Rept. 94-840 (Washington: GPO, 1976), p. 4.

[57] U.S. Small Business Administration, "FY2013 Congressional Budget Justification and FY2011 Annual Performance Report," Washington, DC, p. 35, http://www.sba.gov/sites /default/files/files/FY%202013%20CBJ%20FY%202011%20APR.pdf.

[58] Surety bonds range in price from 0.5% to 2% of the contract price. By dividing the total amount of premiums issued each year by the private sector (about $3.1 billion annually in recent years) by .005 and .02 provides a range for the value of those contracts ($155 billion to $620 billion). Premium data from Surety Information Office, "Contract Surety Bonds, Understanding Today's Market, 2010," Washington, DC, http://www.sio.org/ppt /pptfiles.html.

[59] U.S. Small Business Administration, Office of Congressional and Legislative Affairs.

[60] Ibid.

[61] Ibid.

[62] Ibid.

[63] Ibid. See also U.S. Small Business Administration, "Prior Approval Surety Companies," Washington, DC, http://www.sba.gov/content/prior-approval-surety-companies; and U.S. Small Business Administration, "Preferred Surety Bond Companies," Washington, DC, http://www.sba.gov/content/preferred-surety-bond-participants.

[64] U.S. Small Business Administration, "Bond Agencies By State," Washington, DC, http://www.sba.gov/sites/default/files/Bond%20Agencies%20by%20State%2027%20April %202011.pdf.

[65] 13 C.F.R. § 126.612; 15 U.S.C. § 657a(b)(2)(A)(i)-(iii) (statutory requirements); 48 C.F.R. § 19.1306(a)(1)-(6) ; and Department of Defense, General Services Administration, and National Aeronautics and Space Administration, "Federal Acquisition Regulation: Inflation Adjustment of Acquisition-Related Thresholds," 75 Federal Register 53129, August 30, 2010.

[66] P.L. 111-5, tit. II, § 508, 123 Stat. 158-59 (Feb. 17, 2009).

[67] Testimony of Jeanne Hulit, Acting Associate Administrator for Capital Access, Small Business Administration, before the Subcommittee on Contracting and Workforce of the Committee on Small Business "Construction Contracting: Barriers to Small Business Participation" (Feb. 9, 2012).

[68] The White House, "Section-by-Section Analysis and Explanation of the American Jobs Act of 2011," Washington, DC, September 12, 2011, http://www.whitehouse.gov/blog/2011/09/12 /president-obama-sends-american-jobs-actcongress.

[69] Report run on February 2, 2012, using the Federal Procurement Data System for contracts awarded by the Department of Defense to small businesses where the product services code assigned to the contract was for construction. (On file with the House Committee on Small Business).

70 13 C.F.R. §115.12.

[71] National Association of Surety Bond Producers and The Surety & Fidelity Association of America, "Revitalizing the SBA Bond Guarantee Program," Washington, DC, May 24, 2010.

[72] 15 U.S.C. § 631(a).

[73] Technically, the goal is for the percentage of contract and subcontract dollars awarded to small businesses owned and controlled by socially and economically disadvantaged individuals. While members of certain racial and ethnic groups are presumed to be socially disadvantaged, they must still establish economic disadvantage. Likewise, individuals who are not members of the presumptive groups may establish social disadvantage. See 15 U.S.C. § 631(f)(1)(C); 13 C.F.R. § 124.103(b)(1).

[74] 15 U.S.C. § 644(g)(1).

[75] 15 U.S.C. § 637(a) (set-asides and sole-source awards for small disadvantaged businesses); 15 U.S.C. § 637(m) (set-asides for women-owned small businesses); 15 U.S.C. § 657f (set-asides and sole-source awards for service-disabled veteran-owned small businesses).

[76] See, e.g., National Iranian American Council, NIAC Submits Petition to the SBA for Group Inclusion of Iranian Americans to the 8(a) Program, Mar. 9, 2005, available at http://www.niacouncil.org/index.php?option=com_content&task=view&id=189&Itemid=29 .

[77] The Alaska Native Claims Settlement Act requires that SBA permit ANC firms to compete as 8(a) firms, but does not permit SBA to place the same affiliation and control rules on these firms that would normally be applied to program participants. See 43 U.S.C. § 1626(e).

[78] Federal Acquisition Regulation; Constitutionality of Federal Contracting Programs for Minority-Owned and Other Small Businesses. Proposed Rule, 76 Fed. Reg. 55849 (Sept. 9, 2011)

[79] See, e.g., Dynalantic Corp. v. U.S. Dep't of Defense, 503 F. Supp. 2d 262 (D.D.C. 2007) (denying parties' motions for summary judgment).

[80] Consolidated Appropriations Act, 2001, P.L. 106-554, tit. VIII, § 811, 114 Stat. 2763A–708 (Dec. 21, 2000).

[81] P.L. 109-461, 120 Stat. 3431 (Dec. 22, 2006) (codified, in part, at 38 U.S.C. §§ 8127-8128).

[82] See http://www.acq.osd.mil/osbp/mentor_protege/).

[83] Wendy Schacht, Congressional Research Service, Small Business Innovation Research (SBIR) Program, CRS Report R96-40.

[84] Ibid.

[85] Ibid.

[86] Ibid.

[87] Joel Johnson in oral testimony before the Panel on February 6, 2012.

[88] 48 C.F.R. § 19.704(a)(1)-(11). Prospective contractors that are not themselves small businesses are generally required to submit a proposed subcontracting plan as part of their bid or offer, and agencies may not find a contractor affirmatively "responsible" for purposes of the award of a federal contract unless it agrees to a plan that is also acceptable to the agency. See 48 C.F.R. § 19.705-4. As an alternative to the plans described here, contractors may establish "master plans" that contain similar elements and are valid for three years, or "commercial plans," which apply to the entire production of commercial items sold by the company or a portion of it, in the case of contractors furnishing commercial items. 48 C.F.R. § 19.704(d).

[89] 15 U.S.C. § 644(o).

[90] Subcontracting Transparency and Reliability Act, H.R. 3893, Title I contains a detailed version of this proposal.

[91] See 31 U.S.C. §§ 3901-3907. The Prompt Payment Act does, however, impose certain obligations to pay subcontractors upon federal construction contractors. See 31 U.S.C. § 3905(b).

[92] There does not appear to be any published federal case in which a small business attempted to assert that it was a third party beneficiary of the subcontracting plan in a government contract. See, e.g., Ralte v. Helen Keller Int'l, Inc., 1998 U.S. App. LEXIS 6573, at *8 n.3 (4th Cir. Mar. 31, 1998) (noting that the plaintiff did not allege that she was a third-party beneficiary of the contract in whose subcontracting plan she was listed).

[93] DFARS 252.903 and 252.906(a) (ii).

[94] 15 U.S.C. § 637(d)(12)(C). The act does not define what constitutes a history of "unjustified failure" to make full or timely payments. However, proposed regulations promulgated under the authority of the act would define this to mean three incidents within a 12 month period. Small Bus. Admin., Small Business Subcontracting: Proposed Rule, 76 Fed. Reg. 61625, 61628 (Oct. 5, 2011).

[95] 76 Fed. Reg. at 61626-32.

[96] Ibid. at 61628.

[97] Ibid.

[98] See Government Accountability Office, Department of Energy: Improved Oversight Could Better Ensure Opportunities for Small Business Subcontracting, GAO-05-459 (May 2005).

[99] 76 Fed. Reg. at 61628.

[100] Government Accountability Office, Agency Contracting Data Should Be More Complete Regarding Subcontracting Opportunities for Small Businesses, GAO-07-205, March, 2007.

[101] Center for Strategic and International Studies, "Defense Contract Trends: U.S. Department of Defense Contract Spending and the Supporting Industrial Base," Washington, DC, p. 35, http://csis.org/files/publication/110506_CSIS_Defense_Contract_Trends-sm2.pdf.

[102] Ibid, p. 31.

[103] Ibid, pp. 31-35.

[104] Linda Hillmer in oral testimony before the Panel on January 17, 2012.

[105] 15 U.S.C. § 632(o)(2). Regulations promulgated under the authority of the 1997 amendments further provide that the definition of bundling "does not apply to a contract that will be awarded and performed entirely outside the United States." 48 C.F.R. § 2.101.

[106] 10 U.S.C. § 2382(c)(1). The 2003 and 2010 amendments also addressed multiple award contracts, which were not explicitly addressed in the 1997 amendments. Compare 15 U.S.C. § 657q(a)(2) (defining "consolidation," in part, as the "use of a solicitation to obtain offers for a single contract or a multiple award contract") and 10 U.S.C. § 2382(c)(1) (same) with 15 U.S.C. § 632(o)(2) (defining "bundling," in part, as consolidating requirements "into a solicitation of offers for a single contract"). However, regulations promulgated under the authority of the 1997 amendments incorporated many types of multiple award contracts within their definition of "single contract." See 13 C.F.R. § 125.2(d)(iii)(A)-(B); 48 C.F.R. § 2.101. Requirements performed overseas are apparently not excluded from the definition of "consolidation" under the 2003 amendments. See 48 C.F.R. § 207.107-2.

[107] See, e.g., Tyler Construction Group v. United States, 570 F.3d 1329, 1335 (Fed. Cir. 2009) ("The statute does not prohibit all bundling of contract requirements, but only 'unnecessary and unjustified' bundling.").

[108] See GSA Federal Procurement Data System-Next Generation (FPDS-NG) Data Element Dictionary 96-99 (v.1.4.2, Sept. 23, 2011), available at https://www.fpds.gov /downloads/Version_1.4.2_specs/FPDSNG_DataDictionary_V1.4.2.pdf. The consolidation threshold noted here does not appear to have been adjusted in light of the adjustments for inflation made to federal acquisition regulations in October 2010. See Ronald W. Reagan National Defense Authorization Act for fiscal year 2005, P.L. 108-375, § 807, 118 Stat. 2010-11 (Oct. 28, 2004); 48 C.F.R. § 207-170-3(a).

[109] These numbers should be treated with caution, given the widely discussed issues regarding the quality of data in FPDS. See generally Government Accountability Office, Improvements Needed to the Federal Procurement Data System-Next Generation, GAO-05-960R (Sept. 27, 2005), available at http://www.gao.gov/new.items/d05960r.pdf.

[110] Federal Procurement Data System generated February 2012, on file with the House Committee on Small Business.

[111] See Contractor Opportunity Protection Act of 2012, H.R. 4081.

[112] See Small Business Protection Act of 2012, H.R. 3987.

[113] See Government Accountability Office, Additional Guidance Needed to Improve Visibility into the Structure and Management of Major Weapon System Subcontracts, GAO-11-61R (Oct. 2010); and Enhancing Small-Business Opportunities in the DOD, RAND Report No. TR 601-1 (2008).

[114] The mandatory performance elements reads: "Support the attainment of established DOD small business goals by considering potential small business contracting opportunities during the acquisition process and by establishing a command or program climate that is responsive to small business concerns. Ensure that small business awareness, outreach and support is incorporated as part of the command's overall mission and establish performance

measures that reflect that commitment. Establish, for acquisitions under the executive's purview, annual goals for awards to small business concerns in each category that has a statutory goal. The goal should not be less than the performance achieved during the preceding fiscal year. Develop a corresponding spend plan that establishes the forecasted performance baseline, based on known procurement actions in the budget that can be used to track and report progress to the USD(AT&L). Establish, for acquisitions under the executive's purview, annual goals for awards to small business concerns in each category that has a statutory goal. The goal should not be less than the performance achieved during the preceding fiscal year. Develop a corresponding spend plan that establishes the forecasted performance baseline, based on known procurement actions in the budget that can be used to track and report progress to the USD(AT&L)."

[115] Jayme A. Sokolow, Ph.D. and R. Dennis Green, "Wright Brothers' 1908 Proposal for a Heavier-Than-Air Flying Machine," Proposal Management, Spring 1999, p. 29.

[116] Based on email exchange with Boeing official, January 25, 2012.

[117] For a discussion on the growth of regulation, see James F. Nagle, History of Government Contracting (The George Washington University, 1999), p. 485.

[118] The Office of Federal Procurement Policy was established by Congress in 1974 to provide overall direction for government-wide procurement policies, regulations and procedures and to promote economy, efficiency, and effectiveness in acquisition processes.

[119] James F. Nagle, History of Government Contracting (The George Washington University, 1999), p. 483.

[120] U.S. Congress, Senate Committee on Armed Services, 103rd Cong., 2nd sess., 1994, 258 (Washington: GPO, 1994), p. 2.

[121] Carl L. Vacketta, Federal Government Contract Overview, http://library.findlaw.com /1999/Jan/1/241470.html.

[122] History of Government Contracting, p. 519.

[123] See History of Government Contracting, p. 485, 510. See also FAIR Institute, The State of Competition: Enhancing Competition and Increasing Innovation Across the Federal Government Supply Chain, Executive Summary of FAIR Institute's Report on Competition, October 29, 2009, p. 2.

[124] Office of Manufacturing and Industrial Base Policy, Annual Industrial Capabilities Report to Congress, September 2011, p. 7. Found at http://www.acq.osd.mil/mibp/docs /annual_ind_cap_rpt_to_congress-2011.pdf.

[125] See http://www.acq.osd.mil/osbp/.

[126] Linda Hillmer in oral testimony before the Panel on January 17, 2012

[127] Pentagon's New Power Panel: Will Shape Strategy, Business Plans. DefenseNews. October 24, 2011.

[128] Department of Defense Directive 5105.79.

[129] Department of Defense, Defense Acquisition Guidebook.

[130] Department of Defense Instruction 5000.02, Operation of the Defense Acquisition System, December 8, 2008.

[131] Ibid.

[132] This data source for this estimate is the Office of the Undersecretary of Defense for Acquisition, Technology and Logistics, Data Mart, as part of the Human Capitol Initiative's First Quarter estimates for FY2012, as of 12/31/11. Accessed online at https://dap.dau.mil/workforce/Documents/FY12-Q1%20DAW%20Snapshot%20Count%20 Matrix-v1(2-16-12).pdf.

[133] Defense Science Board Task Force in Improvements to Services Contracting, March 2011, http://www.acq.osd.mil/dsb/reports/ADA550491.pdf.

[134] Dr. Allan Burman in oral testimony before the Panel on February 6, 2012.

[135] Peters, Katherine McIntire. Gates Announces Major Cuts At Defense. Government Executive, August 9, 2010.

[136] David Berteau in oral testimony before the Panel on November 18, 2011.

[137] U.S. Government Accountability Office, Acquisition Workforce: DOD's Efforts to Rebuild Capacity Have Shown Some Progress. GAO-12-323-T, November 16, 2011, p. 2.

[138] Jacques S. Gansler, Democracy's Arsenal Creating a Twenty-First Century Defense Industry, 2011, p. 237

[139] U.S. Congress. House Committee on Oversight and Government Reform, Subcommittee on Technology, Information Policy, Intergovernmental Relations and Procurement Reform. Hearing on Government Acquisition Personnel Training. Testimony of John Hutton, Director, Acquisition and Sourcing Management, GAO, November16, 2011.

[140] P.L. 110-181, the National Defense Authorization Act for Fiscal Year 2008, Sec 852.

[141] Ibid.

[142] Office of the Secretary of Defense, Legislative Liaison Office, provided to the Panel on March 15, 2012.

[143] U.S. Government Accountability Office. Acquisition Workforce: DOD's Efforts to Rebuild Capacity Have Shown Some Progress. GAO-12-323-T, November 16, 2011, p. 4.

[144] The Defense Acquisition Workforce Improvement Act (DAWIA) is the basis for nearly all of DOD's education, training and career development programs for the acquisition workforce. Congress enacted DAWIA in the National Defense Authorization Act for Fiscal Year 1991.[144] It is codified in Chapter 87, Title 10 of the U.S. Code, and has been amended several times since enactment. For further information, see the Defense Acquisition Strategic Workforce Plan, Report to Congress. April 27, 2010.

[145] Ibid.

[146] Ibid.

[147] U.S. Government Accountability Office, Acquisition Workforce: DOD's Efforts to Rebuild Capacity Have Shown Some Progress. GAO-12-323-T, November 16, 2011, p. 4.

[148] U.S. Congress. Human Capital Management: High-Risk Area for the Department of Defense. Testimony of Keith Charles, Director, Human Capital Initiatives, Office of the Undersecretary of Defense for Acquisition, Technology and Logistics, before the House Armed Services Committee, July 14, 2012. Also, Section 874 of P.L. 111-383 requires the Secretary of Defense to establish recertification, training, continuing education, and periodic renewal of individual requirements for the various skill competencies within the defense acquisition workforce. P.L. 111-383, the Ike Skelton National Defense Authorization Act for Fiscal Year 2011, was signed into law on January 7, 2011.

[149] See Department of Defense Defense-wide Operation and Maintenance Budget Justification Book for Fiscal Year 2013 at http://comptroller.defense.gov/defbudget/fy2013 /budget_justification/index.html.

[150] Ibid.

[151] Meeting of the Defense Business Board, July 17, 2008. Accessed online at http://dbb.defense.gov/pdf/07_17_08_minutes.pdf.

[152] See proposed language in H.R. 3890.

[153] Stephen Cross in oral testimony before the Panel on January 23, 2012.

[154] National Research Council (2007), Innovation Policies for the 21st Century, pg. 3.

[155] Ibid, pg. xiii.

[156] "We must continue to maintain our margin of technological superiority." Chairman of the Joint Chiefs of Staff, The National Military Strategy of the United States of America 2011: Redefining America's Military Leadership, Washington, DC, February 8, 2011, p. 18.

[157] In the 1990s DOD reduced the reviews and reporting requirements associated with getting approval or credit for IR&D projects, allowing firms to pick which technologies to pursue as long as they were of potential interest to DOD. In March 2011, DOD put out a proposed rule that would require firms that had greater than $50,000 in IR&D to report their projects to the Defense Technical Information Center and to update their reports annually.

[158] Stephen Cross in oral testimony before the Panel on January 23, 2012.

[159] Peck, M. J. and Frederic M. Scherer. (1962) The Weapons Acquisition Process: An Economic Analysis, Division of Research Graduate School of Business Administration, Harvard University, Boston, pg. 12.

[160] U.S. Congress, House Armed Services, Emerging Threats and Capabilities, Statement Testimony, The Honorable Zachary J. Lemnios, Assistant Secretary of Defense for Research and Engineering, 112th Cong., 1st sess., March 1, 2011.

[161] David Berteau in testimony before the Panel on November 18, 2011.

[162] Statement on the Defense Strategic Guidance Delivered by Secretary of Defense Leon E. Panetta. The Press Briefing Room, The Pentagon, Thursday January 5, 2012.

[163] Trey Hodgkins in oral testimony before the Panel on September 20, 2011.

[164] Science and Technology Policy Institute, Technology Transfer and Commercialization Landscape of the Federal Laboratories (April 2011): https://www.ida.org/upload/stpi/pdfs/p-4728nsfinal508compliantfedlabttcreport.pdf.

[165] Norman Winarsky in oral testimony before the Panel on January 23, 2012.

[166] National Aeronautics and Space Administration and United States Department of Agriculture had technology transfer authorities prior to 1980.

[167] 15 U.S.C. § 3701.

[168] Ibid.

[169] Formally known as the Patent and Trademark Act Amendments of 1980.

[170] 15 U.S.C. § 3701.

[171] Ibid.

[172] U.S. Government Accountability Office, Warfighter Support: DOD's Urgent Needs Processes Need a More Comprehensive Approach and Evaluation for Potential Consolidation, March 2011.

[173] Pierre Chao in oral testimony before the Panel on October 24, 2011.

[174] The Army, Air Force, Navy and Marine Corps, Defense Logistics Agency and U.S. Special Operations Command each have unique supplements.

[175] Far subpart 1.102(a).

[176] Carl L. Vacketta, Federal Government Contract Overview, http://library.findlaw.com/1999/Jan/1/241470.html.

[177] Ibid. See also Grant Thornton, 16th Annual Government Contractor Industry Survey Highlights Book, Industry Survey Highlights 2010, p. 7.

[178] Trey Hodgkins in testimony before the Panel on September 20, 2011.

[179] The State of Competition: Enhancing Competition and Increasing Innovation Across the Federal Government Supply Chain, Executive Summary of FAIR Institute's Report on Competition, October 29, 2009, p. 3. and attached presentation, p. 6.

[180] See "Government Contracts: The Fraud of the Contractors," New York Times, February 6, 1862, p. 2.

[181] U.S. Congress, House Committee on Armed Services, The Department of Defense at High Risk: The Recommendations of the Chief Management Officer on Acquisition Reform and Related High Risk Areas, 111th Cong., 1st sess., May 6, 2009, p. 8.

[182] Based on a Congressional Research Service review of the NDAAs for fiscal year 2008-2012.

[183] Defense Acquisition Reform: An Elusive Goal - 1960 to 2010, J. Ronald Fox, Working Paper 11-120, 2011 Appendix B. With permission from the author.

[184] Pierre Chao, in testimony before the Panel on October 24, 2011.

[185] http://www.acq.osd.mil/dsb/reports/ADA499566.pdf.

[186] David Berteau, in testimony before the Panel on November 18, 2011.

[187] U.S.Congress. On the Frontlines in the Acquisition Workforce's Battle Against Taxpayer Waste. Testimony by John Hutton, Director, Acquisition and Sourcing Management, U.S. Government Accountability Office, before the House Oversight and Government Reform Subcommittee on Technology, Information Policy, Intergovernmental Relations and Procurement Reform, November 16, 2011.

[188] A patent is a grant made by the government conferring on the creator of an invention the exclusive rights to make, use, or sell that invention for a fixed period of time (generally twenty years).

[189] The FAR defines "technical data" as: recorded information ... of a scientific or technical nature (including computer databases and computer software documentation). This term does not include computer software or financial, administrative, cost or pricing, or management data or other information incidental to contract administration. The term includes recorded information of a scientific or technical nature that is included in computer databases.

48 C.F.R. § 2.101. There is a related category of rights in computer software and computer software documentation, which is generally treated the same as rights in technical data, although "limited rights," discussed below, are known as "restricted rights" when computer software and computer software documentation is involved. See, e.g., 48 C.F.R. Subpart 227.72 (rights in computer software and computer software documentation generally); 48 C.F.R. § 227.7204 (rights in computer software and software documentation under SBIR contracts).

[190] When the invention is made in the performance of work under a government contract, the contractor generally may claim ownership of the patent after making the requisite disclosure of the discovery to the government unless certain narrow exceptions apply (e.g., the contractor is not located in the United States). 48 C.F.R. § 27.302(b)(1)-(2).

[191] 48 C.F.R. § 27.302(c).

[192] The government's unlicensed use of a patented invention is viewed as "a taking," under eminent domain, of a license under the patent. See, e.g., Marconi Wireless Telegraph Co. of Am. v. Simon, 246 U.S. 46 (1918). Section 1498 of Title 28 of the United States Code authorizes patent holders to sue the U.S. government in the U.S. Court of Federal Claims to recover the "reasonable and entire compensation" for the government's unlicensed use of their patents in such circumstances.

[193] 48 C.F.R. § 52.227-1. There are two alternate versions of this clause, whose use is required in particular circumstances. Id.

[194] Id. See also 48 C.F.R. § 27.201-1(a) ("[T]here is no direct cause of action against a contractor that is infringing a patent or copyright with the authorization or consent of the Government (e.g., while performing a contract)."). However, the contract could include a "Patent Indemnity" clause obligating the contractor to indemnify the government for any liability

that the government incurs as a result of the contractor's infringement. See 48 C.F.R. § 52.227-3.

[195] Companies Raise Intellectual Property Protection Issues, supra note 3. The FAR is arguably unclear as to whether use of the "Authorization and Consent" clause is required. While Subpart 27.201-2(b) indicates that the "Government may expressly authorize and consent to a contractor's use or manufacture of inventions covered by U.S. patents," Subpart 27.201-2(a) suggests that contracting officers must include the standard clause, or one of its alternates, unless using simplified acquisition procedures, or complete performance and delivery are outside the United States.

[196] A "noncommercial item" is any item that does not fall within the FAR's definition of "commercial item." See 48 C.F.R. § 2.101 (defining "commercial item" to include "(1) [a]ny item, other than real property, that is of a type customarily used by the general public or by non-governmental entities for purposes other than governmental purposes, and (i) [h]as been sold, leased, or licensed to the general public; or (ii) [h]as been offered for sale, lease, or license to the general public.").

[197] Based upon the DFARS) as it existed on January 4, 2012. However, it is important to note that the DFARS has not yet been updated to reflect certain changes pertaining to rights in technical data made by Section 824 of the National Defense Authorization Act for fiscal year 2011 or Section 815 of the National Defense Authorization Act for fiscal year 2012.

[198] 48 C.F.R. § 227.7102-1(b)(2).

[199] "Form, fit, or function data" generally means "data relating to items, components, or processes that are sufficient to enable physical and functional interchangeability, and data identifying source, size, configuration, mating and attachment characteristics, functional characteristics, and performance requirements." 48 C.F.R. § 27.401.

[200] 48 C.F.R. § 227.7102-1(a)-(b).

[201] 48 C.F.R. § 227.7103-5(b)(4). See also 48 C.F.R. § 227.7103-5(c)(2)-(4) (prohibiting the use, release, or disclosure outside the government of data in which the government has limited rights unless certain conditions are met).

[202] See, e.g., Louis D. Victorino, Frankenstein's Monster: Data Rights Changes Adopted in the National Defense Authorization Act for Fiscal Year 2011, Government Contracting. & International Trade Blog, Apr. 18, 2011, available at http://www.governmentcontracts lawblog.com/2011/04/articles/technical-data/.frankensteins-monster-data-rights-changes-adopted-in-the-national-defense-authorization-act-for-fiscal-year-2011.

[203] Companies Raise Intellectual Property Protection Issues, supra note 3.

[204] Spectrum Sciences & Software, Inc. v. United States, 98 Fed. Cl 8 (2011).

[205] This reference is not available in the original document.

[206] 48 C.F.R. § 227.7103-17(b) (emphasis added). The language regarding the "completion of the project under which the data were generated" is significant because SBIR projects can have multiple phases lasting over several years.

[207] 48 C.F.R. § 227.7104(b)(1)-(3).

[208] See, e.g., IP Protection and Sole-Source Status: The SBIR Awardee's Strategic Advantage, 2005, available at http://www.sbircoach.com/files/Phase%20III%20IP%20Presentation.pdf. The only version of the SBIR Policy Directive that could be located on the SBA Website is undated, but includes at least some of the content reportedly in the 2002 directive. See http://archive.sba.gov/idc/groups/public/documents/sba_program_office/sbir_policy_directi ve.pdf.

[209] Companies Raise Intellectual Property Protection Issues, supra note 3.

[210] See, e.g., Reauthorization of the Small Business Innovation Research Program: How to Address the Valley of Death, the Role of Venture Capital, and Data Rights: Roundtable Before the Committee on Small Business and Entrepreneurship, U.S. Senate.

[211] Ibid, p. 52.

[212] Joel L. Johnson in oral testimony before the Panel on February 6, 2012.

[213] The Boeing Company has to pay $15 million dollars because an electric part was used in the export of a commercial aircraft and which was also used in a guided missile which was under export control.

[214] Senate Report 112-043 – Intelligence Authorization Act for Fiscal Year 2012. ("The nation's technological edge, especially in strategic defense systems, aeronautical and missile technologies, nuclear, space, and cyberspace programs must be protected. The Intelligence Community, in cooperation with other agencies of the Federal Government, is well positioned to determine the threat that any potential technological export might pose to U.S. systems, U.S. technological dominance, or U.S. national security.")

[215] Congressional Research Service, China: Possible Missile Technology Transfers Under U.S. Satellite Export Policy — Actions and Chronology, Report 98-485, updated October 6, 2003.

[216] Congressional Research Service, The Buy American Act: Requiring Government Procurements to Come from Domestic Sources, John R. Luckey, March 13, 2009.

[217] 10 U.S.C. 2533a. Requirement to Buy Certain Items from American Sources; Exceptions.

[218] Congressional Research Service, The Jones Act: An Overview, John F. Fritelli, July 8, 2003.

[219] See Grant Thornton, 16th Annual Government Contractor Industry Survey Highlights Book, Industry Survey Highlights 2010, pp 13-15.

[220] U.S. Government Accountability Office, Efforts to Reduce the Cost to Manage and Oversee DoD Contracts, GAO/NSAID96-106, April 18, 1996, http://www.gao.gov/archive/1996 /ns96106.pdf.

[221] Form 10K Annual Report Filing to the Securities and Exchange Commission for Fiscal Year 2010, Raytheon Company, February 23, 2011, p.16-17.

[222] Harvey Sapolsky, "Let's Skip Acquisition Reform This Time," DefenseNews, February 9, 2009, p. 29.

[223] See Thomas Christie, "Sound Policy, Awful Execution," DefenseNews, December 15, 2008, p. 53. Thomas Miller, "Rearranging Deck Chairs on the Titanic: Why Does Acquisition Reform Never Work?," Defense AT&L.

[224] Selected Acquisition Reports are DOD documents describing DOD acquisition programs.

[225] For example, see Obaid Yousossi, Mark V. Arena, and Robert S. Leonard, et al., Is Weapon System Cost Growth Increasing?, RAND, Santa Monica, CA, 2007 Accountability Office, Tools to Prevent Defense Department Cost Overruns, 112th Cong., 1st sess., March 29, 2011.

[226] See Obaid Yousossi, Mark V. Arena, and Robert S. Leonard, et al., Is Weapon System Cost Growth Increasing?, RAND, Santa Monica, CA, 2007.

[227] Congressional Research Service, The Nunn-McCurdy Act: Background, Analysis, and Issues for Congress, Report R41293, January 31, 2012.

[228] U.S. Government Accountability Office, DOD Needs to Monitor and Assess Corrective Actions Resulting from Its Corrosion Study of the F-35 Joint Strike Fighter, December 16, 2010.

[229] Ibid, p. 16.

[230] 22 USC § 2778 - Control of Arms Exports and Imports. ("(f) Periodic review of items on Munitions List; exemptions")

[231] Section 1073 of the National Defense Authorization Act for Fiscal Year 2011, Public 111-383, and the 2011 Defense Appropriation Act provide the Department of Defense with authorities and funds to facilitate the rapid insertion of innovative technologies into military systems or programs that meet critical national security needs.

[232] Phase III is the period during which Phase II innovation moves from the laboratory into the marketplace. No SBIR funds support this phase. The small business must find funding in the private sector or other non-SBIR federal agency funding. To commercialize their product, small businesses are expected to garner additional funds from private investors, the capital markets, or from the agency that made the initial award. The availability of additional funds and the need to complete rigorous testing and certification requirements can pose significant challenges for new technologies and products developed under SBIR awards.

[233] Phase I grants essentially fund a feasibility study in which award winners undertake a limited amount of research aimed at establishing an idea's scientific and commercial promise. The 2012 reauthorization legislation standardized Phase I grants at $150,000.

[234] Phase II grants are larger—typically about $1,000,000 (previous to the fiscal year 12 reauthorization it was $750,000) at DOD—and fund more extensive R&D to develop the scientific and technical merit and the feasibility of research ideas.

[235] The Marine Forces Pacific Experimentation Center (MEC) is an operationally focused center that helps the technical community help the Warfighter. The MEC provides venues, tool sets, and coordination planning and execution for exercises, experiments, demonstrations, and military utility assessments. The MEC facilitates access between the technical community and the Warfighter allowing the Warfighter to provide operational feedback to the Science and Technology and Research and Development process. This feedback allows the technical community to better understand the Warfighters'warfighters' needs and is obtained with minimal intrusion on the operational forces. The MEC also provides theater security cooperation support in the form of Science and Technology collaboration with Thailand, the Philippines, Singapore, Indonesia and potentially other PACOM areas of responsibility.

[236] "Valley of Death," has come to describe the challenging transition when a developing technology is deemed promising, but too new to validate its commercial potential and thereby attract the capital necessary for its development. Lacking the capital to develop an idea sufficiently to attract investors, many promising ideas and firms perish.

In: Challenges to the Defense Industrial Base ISBN: 978-1-62618-141-0
Editor: Leena Nwosu

Chapter 2

INDUSTRIAL BASE: U.S. TACTICAL WHEELED VEHICLE MANUFACTURERS FACE PERIOD OF UNCERTAINTY AS DOD PURCHASES DECLINE AND FOREIGN SALES POTENTIAL REMAINS UNKNOWN*

United States Government Accountability Office

WHY GAO DID THIS STUDY

DOD's need for TWVs dramatically increased in response to operational demands and threats experienced in Afghanistan and Iraq. TWVs primarily transport cargo and personnel in the field and include the High Mobility Multipurpose Wheeled and Mine Resistant Ambush Protected vehicles. The U.S. TWV industrial base, which includes manufacturers and suppliers of major subsystems, increased production to meet DOD's wartime requirements. That base now faces uncertainties as DOD's budget declines and operational requirements for these vehicles decrease. In addition to sales to DOD, U.S. manufacturers sell vehicles to foreign governments.

The Senate Armed Services Committee Report on a bill for the National Defense Authorization Act for Fiscal Year 2012 directed GAO to (1) describe

---

* This is an edited, reformatted and augmented version of United States Government Accountability Office, Publication No. GAO-12-859, dated September 2012.

the composition of the U.S. TWV industrial base, (2) determine how many U.S. manufactured TWVs were purchased by foreign governments from fiscal years 2007 through 2011, and (3) identify factors perceived as affecting foreign governments' decisions to purchase these vehicles. GAO analyzed data from DOD on U.S. and foreign government TWV purchases, as well as sales data from the four primary U.S. TWV manufacturers. GAO also collected data from five foreign governments, including those that did and did not purchase U.S. TWVs.

What GAO Recommends

GAO is not making recommendations in this report. DOD, the Department of State, and two manufacturers provided technical or clarifying comments on a report draft that were incorporated as appropriate.

What GAO Found

The U.S. tactical wheeled vehicle (TWV) industrial base includes seven manufacturers that utilize common suppliers of major subsystems, such as engines and armor. Four of these manufacturers reported that their reliance on sales to the Department of Defense (DOD) varies, in part, as they also produce commercial vehicles or parts. Collectively, the seven manufacturers supplied DOD with over 158,000 TWVs to meet wartime needs from fiscal years 2007 through 2011. DOD, however, plans to return to pre-war purchasing levels, buying about 8,000 TWVs over the next several years, in part, due to fewer requirements.

Almost 28,000 U.S.-manufactured TWVs were purchased for use by foreign governments from fiscal years 2007 through 2011. Approximately 92 percent of these vehicles were paid for using U.S. security assistance funds provided to foreign governments. Iraq and Afghanistan were the largest recipients of such assistance, but officials stated that DOD does not plan to continue funding TWV purchases for these countries. While sales to foreign governments are unlikely to offset reductions in DOD purchases, manufacturers reported that foreign sales are becoming an increasingly important part of their revenue stream.

Sales of U.S.-manufactured TWVs to foreign governments may be affected by multiple interrelated factors, including the availability of used DOD vehicles for sale, foreign competition, differing vehicle requirements, and concerns associated with U.S. arms transfer control regimes. U.S. manufacturers said sales of used Army TWVs to foreign governments could affect their ability to sell new vehicles. U.S. manufacturers and foreign governments also identified a number of non-U.S. manufacturers that produce TWVs that meet foreign governments' requirements, such as right-side drive vehicles. While U.S. manufacturers can produce vehicles that meet these requirements, vehicles they produced for DOD generally have not. Finally, manufacturers and foreign officials had mixed views on how the U.S. arms transfer control regimes may affect foreign governments' decisions to purchase U.S. vehicles. U.S. manufacturers and foreign officials expressed concerns with processing times and U.S. end-use restrictions, but foreign officials also said that such concerns have not been a determining factor when purchasing TWVs that meet their requirements.

ABBREVIATIONS

AECA	Arms Export Control Act
DCS	direct commercial sales
DOD	Department of Defense
FMS	Foreign Military Sales
FMTV	Family of Medium Tactical Vehicles
HMMWV	High Mobility Multi-purpose Wheeled Vehicle
JLTV	Joint Light Tactical Vehicle
MRAP	Mine Resistant Ambush Protected
TWV	tactical wheeled vehicle
USML	United States Munitions List

September 13, 2012
Congressional Committees

Tactical wheeled vehicles (TWV) have played a critical role in supporting the warfighter, as most recently demonstrated in Afghanistan and Iraq. TWVs are primarily designed for use by forces in the field to transport cargo and personnel and are capable of operating on primary and secondary roads, trails, and cross-country terrain. The Department of Defense's (DOD) need for

TWVs increased dramatically in response to the operational demands and threats, such as improvised explosive devices, experienced by U.S. forces during Operation Enduring Freedom and Operation Iraqi Freedom. For example, DOD's procurements for two types of TWVs—the High Mobility Multi-purpose Wheeled Vehicle (HMMWV) and the Family of Medium Tactical Vehicles (FMTV)— increased from 5,500 in 2002 to 27,350 vehicles in 2006. To meet these requirements, the U.S. TWV industrial base, which includes vehicle manufacturers and their suppliers of major subsystems and parts, increased vehicle production. However, that industrial base now faces a period of uncertainty as requirements for these vehicles decrease with the withdrawal of U.S. forces from Iraq and the planned drawdown in Afghanistan, along with declines in DOD's budget.

In addition to sales to DOD, the U.S. TWV industrial base also sells vehicles to foreign governments. The sale of U.S. arms and dual-use items, including TWVs, to friendly nations and allies is an integral component of both U.S. national security and foreign policy.[1] The Arms Export Control Act (AECA) authorizes the President to control the export and import of arms.[2] The AECA authorizes the U.S. government to sell arms to foreign governments through government-to-government agreements as part of the Foreign Military Sales (FMS) program.[3] It also authorizes the issuance of export licenses for U.S. manufacturers to sell arms directly to eligible foreign governments, known as direct commercial sales (DCS). Regardless of the method of sale, each FMS request or DCS license application is reviewed by officials from DOD and the Department of State (State) to ensure the sale would not result in harm to U.S. interests and is consistent with national security and foreign policies.

In response to directions in the Senate Armed Services Committee Report[4] on a bill for the National Defense Authorization Act for Fiscal Year 2012, we (1) described the composition of the U.S. TWV industrial base and the extent of manufacturers' reliance on sales to DOD, (2) determined how many U.S. manufactured TWVs foreign governments purchased in fiscal years 2007 through 2011, and (3) identified factors that are perceived as affecting foreign governments' decisions to purchase these vehicles.

To describe the composition of the U.S. TWV industrial base, we obtained, reviewed, and discussed with DOD officials service-level TWV strategy, acquisition planning, budget, and program office documents. We identified the new TWVs acquired by DOD in fiscal years 2007 through 2011 and new TWVs anticipated for acquisition in fiscal years 2012 through 2017—the period projected in current budget documents. We identified vehicle manufacturers that produced those vehicles, either fully or partially, within the United States and consider them part of the U.S. TWV industrial base for the purposes of this review. We obtained past and planned acquisition data for these TWVs from the Army TACOM Life Cycle Management Command, Marine Corps Systems Command, the Mine Resistant Ambush Protected Joint Program Office, the Air Force Air Logistics Center, and the Navy Expeditionary Program Office. We also obtained data from and interviewed the four U.S. manufacturers that produced approximately 92 percent of all TWVs purchased by DOD in fiscal years 2007 through 2011 regarding their reliance on DOD sales and their supplier base. While we did not obtain data from the other three manufacturers on their reliance and supplier base, the DOD TWV procurement data we obtained include their sales to DOD in fiscal years 2007 through 2011. We assessed the reliability of the data reported by DOD and manufacturers through interviews with knowledgeable officials and electronic data testing for missing data, outliers, and obvious errors and determined the data to be sufficiently reliable for our purposes.

To determine how many U.S. TWVs were purchased by foreign governments, we obtained FMS and DCS data for fiscal years 2007 through 2011. We obtained FMS data from the Army TACOM Life Cycle Management Command's Security Assistance Management Directorate and the Marine Corps Systems Command's International Programs Directorate as they directly facilitate the FMS process for TWVs within DOD. We obtained DCS data from the four TWV manufacturers included in our scope. The FMS data we obtained from DOD contain foreign sales data for the three remaining TWV manufacturers, but we did not collect DCS data from them. We limited FMS and DCS data to vehicles that were purchased by DOD in fiscal years 2007 through 2011 and also sold to foreign governments. We assessed the reliability of FMS data reported by DOD by interviewing officials knowledgeable about the data and crosschecking the data with FMS records from the four manufacturers we met with. We assessed the reliability of DCS data reported by the four manufacturers we met with by interviewing officials knowledgeable about the data and performing electronic testing. Based on those efforts, we determined that the FMS and DCS data reported to us were sufficiently reliable for our purposes.

To identify factors perceived as affecting foreign government decisions to purchase U.S. TWVs, we interviewed representatives from the four U.S. TWV manufacturers included in our scope, foreign government officials from five selected countries, and DOD, State, and Department of Commerce (Commerce) officials. We interviewed officials from three countries that were among the biggest buyers of U.S. TWVs through the FMS program in fiscal years 2007 through 2011 or DCS in fiscal years 2008 through 2011. We also interviewed officials from two allied countries that did not purchase U.S. TWVs through the FMS program in fiscal years 2007 through 2011 or DCS in fiscal years 2008 through 2011 to gain additional perspectives. Some of the five countries were in the process of purchasing TWVs. Collectively, these officials provided information on foreign manufacturers of TWVs and foreign TWV requirements. Because the countries provided us with information on the size and composition of their TWV fleets, which some regard as sensitive, we are not identifying the countries that participated in our review by name. To determine how the U.S. arms transfer control regimes may affect such sales to foreign governments, we also reviewed the AECA, Export Administration Act of 1979, International Traffic in Arms Regulations, DOD's Security Assistance Management Manual, and related U.S. guidance. To determine how many U.S. TWV sales to foreign governments were denied, we interviewed officials from State's Directorate of Defense Trade Controls and Regional Security and Arms Transfer offices and Commerce's Bureau of Industry and Security. We also reviewed available State export licensing data for fiscal years 2008 through 2011 and Commerce's data for fiscal years 2007 through 2011.[6]

We conducted this performance audit from October 2011 to September 2012 in accordance with generally accepted government auditing standards. Those standards require that we plan and perform the audit to obtain sufficient, appropriate evidence to provide a reasonable basis for our findings and conclusions based on our audit objectives. We believe that the evidence obtained provides a reasonable basis for our findings and conclusions based on our audit objectives.

Background

Tactical Wheeled Vehicles

DOD and the military services classify TWVs by weight or payload capacity into three categories—light, medium, and heavy—although the definitions of each class vary among the services. Each class generally includes multiple variants or models built on a common chassis. For example, the Army's FMTV consists of 2.5- and 5-ton capacity trucks, each with the same chassis and includes cargo, tractor, van, wrecker, and dump truck variants. Table 1 lists the TWVs acquired by the military services over five fiscal years, fiscal years 2007 through 2011.

Table 1. DOD TWVs by Type Acquired, Fiscal Years 2007 through 2011

Vehicle class	Vehicle family	Number of variants	Military service
Light	High Mobility Multi-purpose Wheeled Vehicle	5	Army, Marine Corps, Air Force, Navy
Medium	Light Medium Tactical Vehicle	3	Army
	Medium Tactical Vehicle	10	Army
	Medium Tactical Vehicle Replacement	11	Marine Corps
Heavy	Heavy Expanded Mobility Tactical Truck	6	Army
	Palletized Load System	1	Army
	M915 Series Line Haul Tractor	2	Army
	Heavy Equipment Transport System	1	Army
	Logistics Vehicle System Replacement	3	Marine Corps
Not applicable[a]	Firefighting Truck	2	Army
	Mine Resistant Ambush Protected vehicle	13	Army, Marine Corps, Air Force, Navy

Source: GAO summary of DOD information.
[a]The military services do not classify these vehicles by class.

Requirements for TWVs have evolved over the last decade, in part, due to the operational threats encountered in Afghanistan and Iraq. TWVs were traditionally viewed as utility vehicles that required little armor because the vehicles operated behind the front lines. However, the tactics used against forces in these countries dictated that vehicles needed more protection. For example, the HMMWV was conceived and designed to support operations in relatively benign environments behind the front line, but it proved to be highly vulnerable to attacks from improvised explosive devices, rocket-propelled grenades, and small arms fire when it was required to operate in urban environments. As a result, DOD identified an urgent operational need for armored tactical vehicles to increase crew protection and mobility of soldiers. Although the initial solution—the Up-Armored HMMWV—provided greater protection, the enemy responded by increasing the size, explosive force, and type of improvised explosive devices, which were capable of penetrating even the most heavily armored vehicles. Consequently, the Mine Resistant Ambush Protected (MRAP) vehicle was approved in 2007 as a rapid acquisition capability. DOD recognized that no single manufacturer could provide all of the vehicles needed to meet requirements quickly enough, so it awarded contracts to multiple manufacturers.

U.S. Arms Transfer Control Regimes

The AECA authorizes the President to control the export of arms, such as TWVs. [7] The authority to promulgate regulations on these items has been delegated to the Secretary of State.[8] State administers arms transfer controls through the International Traffic in Arms Regulations and designates, with the concurrence of DOD, the articles and services deemed to be arms.[9] These arms constitute the United States Munitions List (USML).[10] DOD's TWVs are generally designated as Category VII (Tanks and Military Vehicles) items on the USML.

Arms, including TWVs, can be sold and exported to foreign governments through the FMS program or DCS. Under the FMS program, the U.S. government procures items on behalf of eligible foreign governments using the same acquisition process used for its own military needs. While State has overall regulatory responsibility for the FMS program and approves such sales of arms to eligible foreign governments, DOD's Defense Security Cooperation Agency administers the program. Alternatively, the DCS process allows foreign governments to directly negotiate with and purchase arms from U.S. manufacturers. For TWVs controlled on the USML, manufacturers must generally apply for an export license to State's Directorate of Defense Trade Controls, which authorizes the export of arms to foreign governments.[11]

State officials assess all arms export requests through the FMS program and DCS license applications against 12 criteria specified in the Conventional Arms Transfer Policy, as summarized in table 2.[12]

DOD officials assess the technical risks of the sensitive or classified electronic equipment associated with the sale of TWVs to foreign governments, including the type of armor, sensors or weapons attached to the vehicle, and any signature information. Aside from these technologies, State and DOD officials said the departments generally consider the technology associated with TWVs comparable to commercially available trucks and do not have any additional policies pertaining to the sale of TWVs to foreign governments.

Table 2. Conventional Arms Transfer Policy Criteria

1	Consistency with U.S. regional stability interests, especially when considering transfers involving power projection capability or introduction of a system which may foster increased tension or contribute to an arms race.
2	The degree to which the transfer supports U.S. strategic and foreign policy interests through increased access and influence, allied burden sharing, and interoperability.
3	The human rights, terrorism and proliferation record of the recipient and the potential for misuse of the export in question.
4	Consistency with international agreements and arms control initiatives.
5	Appropriateness of the transfer in responding to legitimate U.S. and recipient security needs.
6	The impact of the proposed transfer on U.S. capabilities and technological advantage, particularly in protecting sensitive software and hardware design, development, manufacturing, and integration knowledge.
7	The degree of protection afforded sensitive technology and potential for unauthorized third-party transfer, as well as in-country diversion to unauthorized uses.
8	The ability of the recipient effectively to field, support, and appropriately employ the requested system in accordance with its intended end-use.
9	The risk of revealing system vulnerabilities and adversely impacting U.S. operational capabilities in the event of compromise.
10	The impact of U.S. industry and the defense industrial base whether the sale is approved or not.
11	The availability of comparable systems from foreign suppliers.
12	The risk of adverse economic, political or social impact within the recipient nation and the degree to which security needs can be addressed by other means.

Source: Department of State.

In accordance with the AECA, recipient countries of arms, including TWVs, must generally agree to a set of U.S. arms transfer conditions, regardless if sold through the FMS program or DCS.[13] The conditions include agreeing to use the items only for intended purposes without modification, not to transfer possession to anyone not an agent of the recipient country without prior written consent of the U.S. government, and to maintain the security of any defense article with substantially the same degree of protection afforded to it by the U.S. government. To ensure compliance with these conditions, recipient countries must permit observation and review by U.S. government representatives on the use and possession of U.S. TWVs and other arms.

While the majority of TWVs that DOD purchases are regulated on the USML, a small number that lack armor, weapons, or equipment that would allow armor or weapons to be mounted are considered to be dual-use items—having both commercial and military applications. These items are controlled under the Export Administration Act of 1979,[14] which established Commerce's authority to control these items through its Export Administration Regulations and Commerce Control List.[15] On the Commerce Control List, DOD's TWVs are generally designated as Category 9 (Propulsion Systems, Space Vehicles, and Related Equipment) items. For DCS of such items, U.S. manufacturers must comply with the Export Administration Regulations to determine if an export license from the Commerce's Bureau of Industry and Security is required.

U.S. TWV Industrial Base Includes a Number of Manufacturers Whose Reliance on Sales to DOD Vary

Numerous Vehicle Manufacturers and Suppliers Comprise the U.S. TWV Industrial Base

The U.S. TWV industrial base includes seven vehicle manufacturers, over 90 major subsystem suppliers, and potentially thousands of parts and component suppliers. Four of the seven manufacturers provided approximately 92 percent of all TWVs purchased by DOD in fiscal years 2007 through 2011. Figure 1 identifies the manufacturers, the vehicles they produced, and the percent of all vehicles purchased by DOD from each manufacturer in fiscal years 2007 through 2011.

Although these manufacturers produced 11 different families of TWVs, which included over 50 vehicle variants, they generally relied on common suppliers for major subsystem components. For example, the manufacturers relied on six or fewer suppliers to provide components, such as engines or tires. In contrast, the manufacturers relied on more than 25 armor suppliers, in part, because there was a shortage of vehicle armor during initial MRAP production. DOD reported that the requirements for armor, in response to the conflicts in Iraq and Afghanistan, provided an opportunity for several suppliers to begin producing armor, which eventually resolved the armor shortage.

TWV manufacturers	TWV families of vehicles	Percent of DOD's TWV purchases
AM General	• HMMWV	36%
Oshkosh	• Light Medium Tactical Vehicle • Medium Tactical Vehicle • Medium Tactical Vehicle Replacement • Heavy Expanded Mobility Tactical Truck • Palletized Load System • Heavy Equipment Transport System • Logistics Vehicle System Replacement • Firefighting Truck • MRAP variant	31%
BAE	• Light Medium Tactical Vehicle • Medium Tactical Vehicle • MRAP variants	19%
Navistar	• MRAP variants	6%
Freightliner	• M915 Series Line Haul Tractor	5%
Force Protection[a]	• MRAP variants	2%
General Dynamics	• MRAP variant	1%

Source: GAO analysis of DOD data.
[a]General Dynamics acquired Force Protection in December 2011.

Figure 1. TWV Manufacturers, Vehicles Produced, and Market Share for DOD TWVs, Fiscal Years 2007 through 2011.

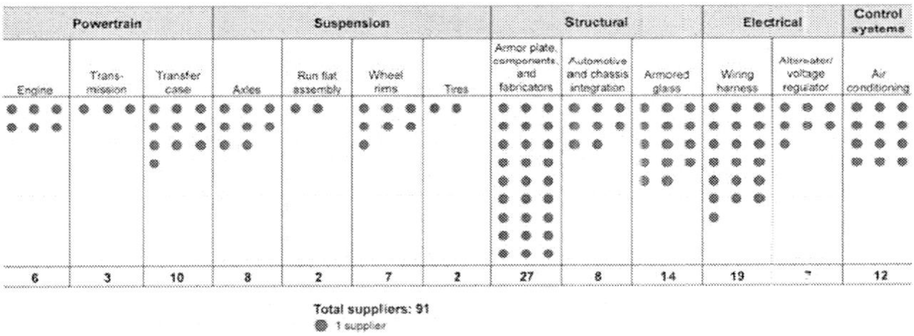

Source: GAO analysis of DOD and manufacturer data.

Note: For the purposes of this analysis, DOD and manufacturers excluded items, such as armaments and mission equipment (communications and situational awareness subsystems), because the military services buy these items for their vehicles and provide them as government furnished equipment. Additionally, 125 suppliers are indicated in the figure but some suppliers provided components for multiple subsystems, reducing the total to 91 suppliers.

Figure 2. Commonality of Suppliers for Major Vehicle Subsystems in DOD TWVs.

In addition to these suppliers, manufacturers we met with reported there were potentially thousands of other companies that produced parts for these vehicles. See figure 2 for more information on the number of suppliers that produced major subsystems on DOD's TWVs.

DOD Plans to Buy Few TWVs over the Next Several Years

DOD purchased over 158,000 TWVs in fiscal years 2007 through 2011 but plans to buy significantly less from now through fiscal year 2017. DOD demands for TWVs increased dramatically in response to the operational demands and threats experienced by U.S. forces during Operation Enduring Freedom and Operation Iraqi Freedom. For example, between fiscal years 1998 through 2001, before these two wars began, Army budget documents indicate plans to purchase approximately 5,000 HMMWVs. After the start of Operation Enduring Freedom, Army budget documents in 2003 reflected an increased requirement for HMMWVs and, at the time, it planned to purchase approximately 23,000 though fiscal year 2009. However, after Operation Iraqi Freedom began, the need for HMMWVs increased further and the Army reported that it ultimately purchased approximately 64,000 between 2003 through 2009.

As U.S. forces began to draw down from the conflicts in Iraq and Afghanistan, DOD's operational requirements for TWVs declined. For example, while DOD bought over 100,000 TWVs in fiscal years 2007 and 2008, DOD plans to purchase less than 1,000 TWVs in fiscal years 2015 and 2016. In all, DOD plans to purchase approximately 8,000 TWVs in fiscal years 2012 through 2017, as shown in figure 3.

Future defense budgets will likely constrain new vehicle purchases and the size of a fleet the military services will be able to sustain. Army officials told us that it would cost approximately $2.5 billion per year to sustain its current fleet of approximately 260,000 TWVs and meet any new TWV requirements. Officials stated, however, that the Army can no longer afford and does not need such a sized fleet, in part, due to budget cuts and potential force structure changes. The Army is re-evaluating how many TWVs it needs and can afford, which will be outlined in a revised TWV strategy. In developing this revised strategy, Army officials recognize that the Army has a relatively young fleet of TWVs, averaging 9 years of age, many of which will be part of its fleet through 2040.

While this revised strategy has not been completed, the Army has already made changes to reduce its TWV costs. For example, in February 2012 the Army reduced the number of FMTVs it planned to purchase by approximately 7,400 vehicles. At that time, the Army also terminated a HMMWV modernization effort, known as the Modernized Expanded Capability Vehicle, which was intended to improve vehicle performance and crew protection on over 5,700 HMMWVs. Officials stated that this effort was terminated, in part, because of DOD-wide funding constraints. Army officials estimate that these actions will result in a total savings of approximately $2.7 billion in fiscal years 2013 through 2017. Furthermore, Army officials stated that the Army plans to reduce the size of its TWV fleet to match force structure requirements. They also stated that, as of July 2012, the Army plans to reduce its total fleet by over 42,000 vehicles. Officials added that more vehicles could be divested depending on any future force structure changes and budget constraints.

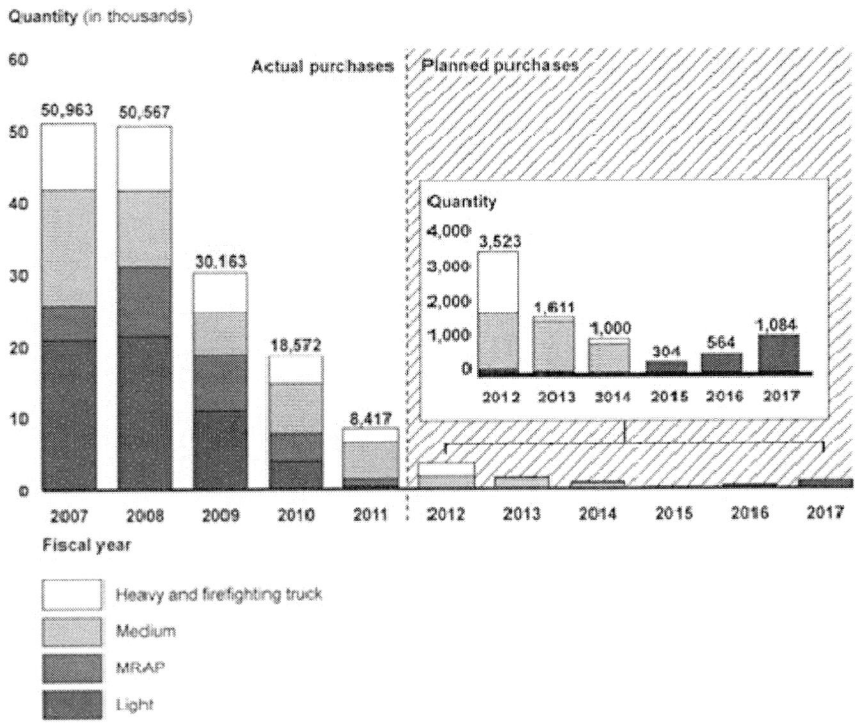

Source: GAO analysis of DOD data and budget documents.

Figure 3. Declining DOD Purchases by TWV Family, Fiscal Years 2007 through 2017.

Despite budget constraints, the industrial base will have some opportunities over the next several years to produce a new TWV for DOD. The Joint Light Tactical Vehicle (JLTV) is a new DOD program, designed to fill the gap between the HMMWV and MRAP by providing near-MRAP level protection while maintaining all-terrain mobility. As we previously reported, the Army and Marine Corps are pursuing a revised developmental approach for JLTV and awarded technology development contracts to three industry teams.[16] The program completed the technology development phase in January 2012. Last month, the Army awarded three contracts for the JLTV's engineering and manufacturing development phase. While production contracts will not be awarded for some time, DOD reports that it plans to purchase approximately 55,000 JLTVs over a 25-year period with full rate production beginning in fiscal year 2018. With production of other TWVs for DOD largely coming to an end in fiscal year 2014, DOD considers the JLTV program to be critical in maintaining an industrial base to supply TWVs to the military.

In addition to new production, the Army and Marine Corps also plan to invest in sustainment efforts that could be completed by the U.S. TWV industrial base. These efforts include restoring or enhancing the combat capability of vehicles that were destroyed or damaged due to combat operations.[17] For example, Marine Corps officials reported that it plans to recapitalize approximately 8,000 HMMWVs beginning in fiscal year 2013. In addition, the Army is in the process of resetting the portion of its FMTV fleet that was deployed through at least fiscal year 2017 as well as recapitalizing some of its heavy TWVs.

Manufacturer Reliance on Sales to DOD Varied

Despite the significant decrease in DOD TWV purchases, the four manufacturers we met with generally reported that these sales remain an important part of their revenue stream.

However, there is a wide range in the degree to which the manufacturers were reliant on DOD in a given year. For example, as shown in table 3, one manufacturer reported that for 2007 its revenue from sales to DOD accounted for 4 percent of its total revenue while another manufacturer reported such revenue was as high as 88 percent, with the other two manufacturers falling within that range.

Table 3. Ranges of TWV Manufacturer Reported Reliance on DOD Sales

	Percent of revenue from DOD sales	
Year	Low	High
2007	4%	88%
2009	26%	72%
2011	14%	73%

Source: GAO analysis of manufacturer data.

Among the four manufacturers, the extent of reliance on revenue from DOD sales varied, in part, because of vehicles sold in the commercial truck and automotive sectors. Aside from producing TWVs, manufacturers produced or assembled commercial vehicles, such as wreckers, fire trucks, school buses, and handicap-accessible taxis, as well as vehicle components, such as engines, transmissions, and suspensions. According to the four manufacturers, their suppliers of TWV major subsystem components generally produced items in the commercial automotive and truck industries. For example, according to manufacturers, suppliers generally produced parts, such as engines, transmissions, axles, and tires for their commercial vehicles in addition to supplying parts for the TWVs they produce. However, vehicle armor, a major TWV component, is primarily a defense-unique item and those suppliers were not typically used in the manufacturers' commercial vehicles.

DOD Has Several Studies under Way to Assess the U.S. TWV Industrial Base

DOD currently has several studies under way to better understand the U.S. TWV industrial base, its capabilities, and how declining DOD sales may affect it. In 2011, DOD's Office of Manufacturing and Industrial Base Policy began a multifaceted review of the U.S. TWV industrial base that includes surveying suppliers, conducting site visits, and paneling experts. The Army's TACOM Life Cycle Management Command also has ongoing studies, including a review to assess the health of the industrial base and others intended to identify its supplier base and any risks associated with sustaining DOD's TWV fleet. Some of the goals of these different studies are to better understand how different vehicle supply chains affect others, identify single point failures in the supply chain, and provide DOD leadership with improved information so they may better tailor future acquisition policies.

TWV Sales to Foreign Governments Were Relatively Few and Generally Purchased with U.S. Funds

U.S. manufacturers sold relatively few TWVs for use by foreign governments in fiscal years 2007 through 2011, when compared to the 158,000 vehicles sold to DOD over that same period. However, most of the manufacturers we met with stated that while sales of TWVs to foreign governments have not equaled those sold to DOD, such sales are becoming an increasingly important source of revenue as DOD purchases fewer vehicles. According to data provided by DOD and the four manufacturers, foreign governments purchased approximately 28,000 TWVs, either through the FMS program or through DCS, in fiscal years 2007 through 2011. In addition to these sales to foreign governments, manufacturers reported they exported approximately 5,000 other TWVs that were different vehicles than those DOD purchased during that time period.

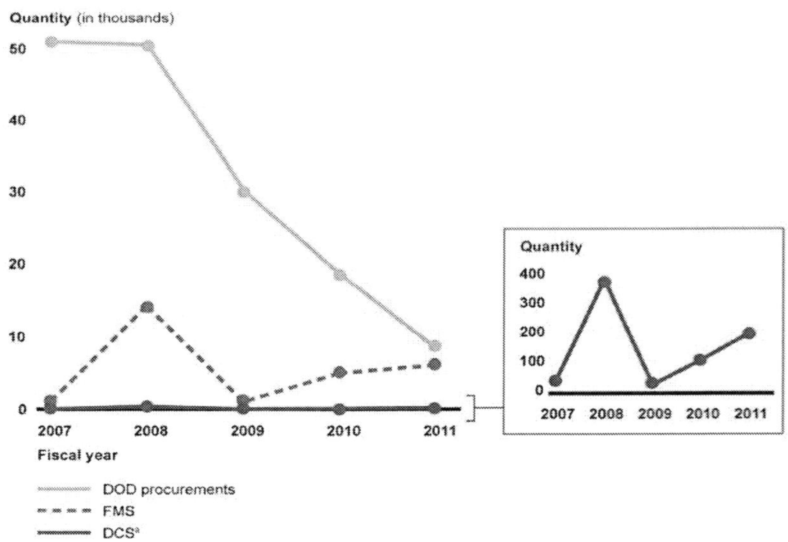

Source: GAO analysis of manufacturer data.

[a]DCS data only reflects data provided by the four manufacturers we met with and is limited to vehicle models purchased by DOD.

Figure 4. Comparison of TWV Sales to DOD and Foreign Governments, Fiscal Years 2007 through 2011.

Nearly all TWVs sold to foreign governments were sold through the FMS program rather than through DCS. DOD reports that about 27,000 TWVs were sold through the FMS program, while the four manufacturers we met with reported that about 700 vehicles were sold through DCS in fiscal years 2007 through 2011.[18] See figure 4 for a comparison of TWVs sold to DOD and to foreign governments through the FMS program and DCS in fiscal years 2007 through 2011.

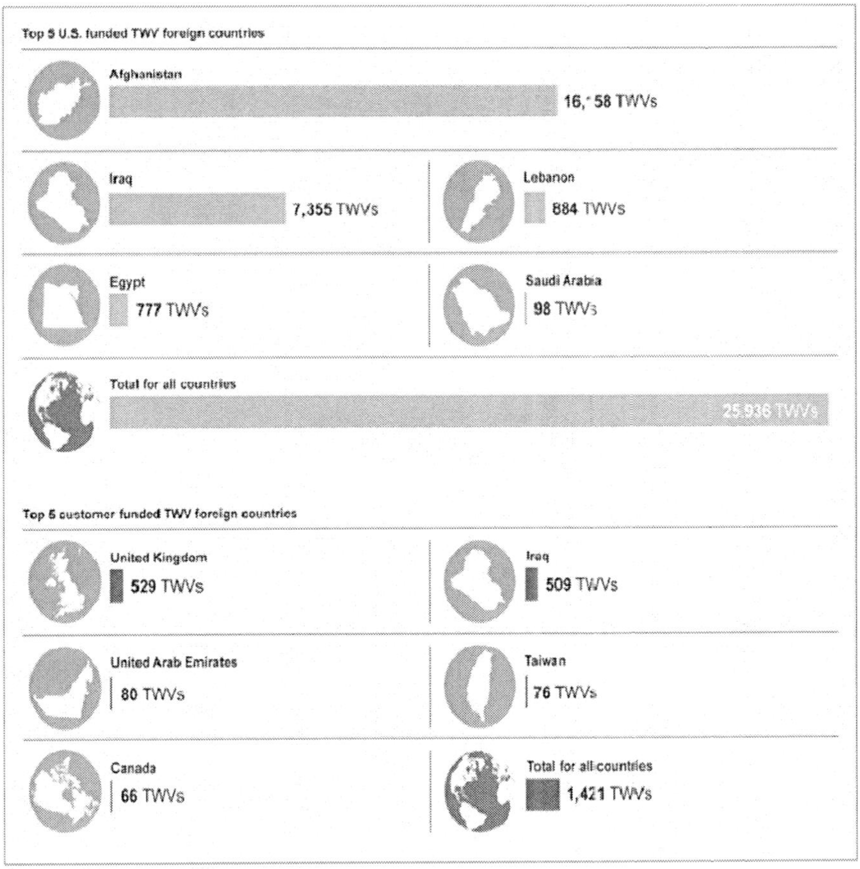

Source: GAO analysis of DOD data (data); Map Resources (images).

Figure 5. Comparison of Funding Types for TWV FMS, Fiscal Years 2007 through 2011.

Approximately 95 percent of TWVs purchased through the FMS program from fiscal year 2007 through 2011 were paid for using U.S. government funding through different security and military assistance programs. The U.S. Congress authorizes and appropriates funds for assistance programs that support activities, such as security, economic, and governance assistance in foreign countries.

Examples of such assistance programs include the Afghanistan Security Forces Fund and Iraq Security Assistance Fund, which were sources of funding for TWVs purchased for Afghanistan and Iraq through the FMS program.

While Afghanistan and Iraq were the largest recipients of U.S. manufactured TWVs through such assistance programs, DOD officials informed us that as the war efforts conclude there, U.S. funding for TWVs for these two countries' security forces has declined and is not planned to continue. In addition, a smaller number of TWVs were sold through the FMS program to countries using their own funds.

Figure 5 identifies the countries that purchased the most U.S. manufactured TWVs with U.S. or their own funds through the FMS program.

U.S. Manufacturers and Foreign Governments Identified Multiple Interrelated Factors That May Affect TWV Foreign Sales

U.S. manufacturers of TWVs and foreign government officials we met with identified a number of interrelated factors that they perceive as affecting whether a foreign government decides to purchase U.S. manufactured TWVs. These included potential future competition from transfers of excess (used) U.S. military TWVs, competition from foreign manufacturers, and differing foreign requirements for TWVs.

In addition, these U.S. manufacturers and foreign government officials expressed mixed views on the effect the U.S. arms transfer control regimes may have on foreign governments' decisions to buy U.S. vehicles. These officials said that processing delays and end-use restrictions can influence foreign governments' decisions to buy U.S. TWVs. Despite these issues, foreign government officials said the U.S. arms transfer control regimes would not adversely affect their decisions to purchase a U.S.-manufactured TWV that best meets their governments' requirements.

Potential Transfers of Used Army TWVs Viewed as Risk by Manufacturers to Future Sales of U.S. TWVs to Foreign Governments

The U.S. manufacturers we met with regard the Army's intent to reduce its TWV fleet size as a risk to their future sales of TWVs to foreign governments. Army officials said it is still assessing its TWV requirements and potential plans to divest over 42,000 vehicles, but they acknowledge that a number of these TWVs could be transferred through the FMS program. The four U.S. manufacturers consider these used vehicles to be a risk to their future sales of U.S. TWVs to foreign governments because foreign governments could be less likely to purchase new vehicles from U.S. manufacturers if the U.S. Army transfers these used vehicles through foreign assistance programs. U.S. manufacturers told us they would like more involvement in DOD's decisions on its plans for these divested vehicles so they may provide input on potential effects on the industrial base. Commerce's Bureau of Industry and Security reviews proposed FMS of divested items to identify effects on the relevant industry. During this review, Commerce provides industry with the opportunity to identify any impacts of the potential FMS on marketing or ongoing sales to the recipient country. When approving these transfers, State and Defense Security Cooperation Agency officials said the U.S. government must also weigh national security and foreign policy concerns, which could outweigh industrial base concerns with transfers of used DOD TWVs to foreign countries. While concerned about the potential for competition from the FMS of these retired vehicles, U.S. manufacturers also view these planned divestitures as a potential to provide repair or upgrade business that could help sustain their production capabilities during a period of low DOD demand. Some manufacturers we met with stated that they would like to purchase DOD's used TWVs, before they are made available to foreign governments, so they may repair or upgrade them and then sell them to foreign governments. DOD is currently reviewing its policies to determine which vehicles, if any, could be sold back to manufacturers. Another manufacturer, while not interested in purchasing the vehicles, expressed interest in providing repair or upgrade services on the used TWVs before they are sold to foreign governments. Defense Security Cooperation Agency officials stated that excess defense articles, such as the used TWVs, are generally made available to foreign governments in "as is" condition and recipient countries are responsible for the cost of any repairs or upgrades they may want to make. They added that in such instances, it could be possible for U.S. manufacturers to perform such services, but it would be at the direction of the purchasing country, not the U.S. government.

Foreign Competition, Different Vehicle Requirements, and Concerns Associated with U.S. Arms Transfer Control Regimes Affect Foreign Governments' Decisions to Purchase U.S. TWVs

Foreign government and manufacturer officials that we interviewed identified a number of TWV manufacturers that compete with U.S. manufacturers for international sales. Examples of foreign manufacturers are shown in table 4.

Table 4. Foreign TWV Manufacturers as Identified by Manufacturers and Foreign Governments

Manufacturer	Headquarters location
Ashok Leyland	India
DAF Truck	The Netherlands
General Dynamics European Land Systems	Spain
IVECO	Italy
Krauss-Maffei Wegmann	Germany
Land Rover	The United Kingdom
Mahindra	India
Mercedes Benz	Germany
Nimr	United Arab Emirates
Patria	Finland
Renault	France
Rheinmetall MAN	Germany
Tata Motors	India
Tatra	Czech Republic
Toyota	Japan
Thales	France
UROVESA	Spain

Source: Officials from foreign governments and U.S. manufacturers..

Officials from two countries that had not purchased U.S. manufactured TWVs explained that their countries have a well established automotive industrial base capable of producing TWVs that meet their governments' needs. While all of the foreign officials we interviewed reported that their countries had no policies that favor their domestic manufacturers, governments that have not purchased U.S. TWVs generally purchased vehicles from domestic manufacturers. For example, foreign officials from one country said that all of their government's TWVs are assembled within its borders.

While all of the competitors to U.S. TWV manufacturers are not headquartered in the purchasing countries, foreign officials reported that many of these companies have established dealer and supplier networks within their countries. Foreign officials reported that these domestic dealer and supplier networks make vehicle sustainment less expensive and more manageable, in part, because it is easier and quicker to obtain replacement parts or have vehicles repaired. In contrast, foreign officials said that U.S. TWV manufacturers do not generally have the same dealer and supplier networks within their countries. They added that this can make maintenance of the U.S. vehicles more expensive, in part, due to the added cost of shipping.

In addition to the number of TWV manufacturer competitors, foreign officials also reported that there is limited foreign demand for TWVs. Foreign officials reported that their governments purchase relatively few TWVs compared to the U.S. government, in part, because their fleet size requirements are much smaller. Foreign officials we interviewed reported TWV fleets that ranged in size from 2 to 9 percent the size of the U.S. Army's fleet. For example, foreign officials from one country stated that their military was in the process of upgrading its entire fleet of approximately 7,500 vehicles, which is less than 3 percent of the size of the U.S. Army's TWV fleet.

Foreign government officials also explained that U.S. manufacturers can generally produce TWVs to meet their governments' requirements, but the vehicles U.S. TWV manufacturers are producing for DOD do not necessarily align with these requirements. Foreign government officials identified the following areas where their governments' requirements differ from those of DOD:

- DOD's TWVs are generally larger than what their government can support. For example, officials from one foreign government reported that its military considered purchasing U.S. manufactured MRAP vehicles but did not have the cargo planes required to transport a vehicle the size and weight of DOD's MRAP vehicles. Instead, according to the official, this country purchased a mine and ambush protected vehicle developed by one of its domestic manufacturers that is smaller and lighter than the DOD's MRAP vehicles and better aligned with its transportation capabilities.

- Their governments do not always require the same level of capabilities afforded by DOD's TWVs and, in some cases, requirements may be met by commercially available vehicles. For example, foreign government officials identified a number of vehicles in their governments' tactical fleets that are based on commercial products from automobile companies such as Jeep and Land Rover.
- Their governments have different automotive or design standards for military vehicles that do not always align with those produced for DOD by U.S. manufacturers. For example, officials from one country said that their military is required to purchase right-side drive vehicles, which are not always supported by U.S. manufacturers. While their military can obtain a waiver to purchase a left-side drive vehicle, this presents training challenges as the majority of the vehicles in its fleet are right-side drive vehicles. Foreign officials said that while U.S. manufacturers are capable of meeting these requirements, foreign competitors may be more familiar with these requirements. Manufacturers that we interviewed said they produce or are developing TWVs to better meet foreign customers' requirements. For example, one U.S. manufacturer said it was developing a right-side drive variant of one of its vehicles and another manufacturer said that it has a line of TWVs for its international customers that better meets those requirements.

U.S. manufacturers and foreign officials expressed mixed views on the effect the U.S. arms transfer control regimes may have on the sale of U.S.-manufactured TWVs to foreign customers. Officials we met with reported that, generally, the U.S. arms transfer control regimes do not inhibit foreign governments from purchasing U.S. manufactured TWVs. Accordingly, we found that once the FMS and DCS process was initiated, no eligible foreign sales or licenses for U.S. TWVs were denied. For example, State officials reported that no countries eligible to participate in the FMS program were denied requests to purchase TWVs in fiscal years 2007 and 2011. Similarly, State DCS license data indicated that no licenses for vehicle purchases were denied from fiscal years 2008 through 2011.

While sales of TWVs to foreign governments are generally approved by the U.S. government once initiated, U.S. manufacturers and foreign officials said that foreign governments may prefer to purchase vehicle manufactured outside the United States, in part, due to the amount of time to process sales and licenses requests and end-use restrictions associated with the U.S. arms transfer control regimes. Specifically, manufacturers said the congressional notification process can result in lengthy delays during the FMS and DCS approval process. The AECA requires notification to Congress between 15 and 45 days in advance of its intent to approve certain DCS licenses or FMS agreements.[19] Preceding the submission of this required statutory notification to the U.S. Congress, State provides Congress with an informal review period that does not have a fixed time period for action. One manufacturer stated that this informal review period, in one case, lasted over a year and, after which, the prospective customer decided to not continue with the purchase. Another manufacturer said that the informal congressional notification process is unpredictable because there is no set time limit for review, making it difficult for the manufacturer to meet delivery commitments to foreign customers. State officials acknowledged that the informal congressional notification period can delay the DCS and FMS process because there is no designated time limit for review. According to State officials, the department established a new tiered review process in early 2012 to address this issue by establishing a time bounded informal review period that is based on the recipient country's relationship with the U.S. government. The formal notification period remains unchanged.

Foreign officials said when TWVs that meet their governments' requirements are available from manufacturers outside the United States, AECA restrictions on third party transfers and end-use administrative requirements associated with U.S. manufactured vehicles could affect their governments purchasing decisions. Foreign officials explained that there are a number of TWV manufacturers outside the United States that can meet their requirements and vehicles sold by those manufacturers do not necessarily come with the same end-use restrictions as U.S. vehicles. For example, the AECA restricts the transfer of arms, including U.S. manufactured, TWVs to a third party without consent of the U.S. government.[20] Some foreign officials said their governments prefer to use private companies, when possible, to make repairs and maintain its TWV fleet because it can reduce costs compared to government repair work. These foreign officials said that U.S. third party transfer restrictions require that their governments obtain permission from the U.S. government before transferring a U.S. TWV to a private company for repairs, which creates an administrative burden. Additionally, foreign governments are required to maintain information on U.S. TWVs' end-use and possession that must be available to U.S. officials when requested to ensure compliance with U.S. end-use regulations. Foreign officials from one country said the maintenance of this information is an administrative burden and will be more difficult to manage as their government tries to reduce its workforce in a limited budget environment. Foreign officials said that TWVs purchased from manufacturers outside of the Untied States are not generally encumbered with these same restrictions and administrative burdens, making maintenance of these vehicles easier and cheaper, in some cases. State officials acknowledged these concerns from foreign governments but said these restrictions play an important role in protecting U.S. national security interests.

Foreign officials reported, however, that the U.S. arms transfer control regimes would not adversely affect their decision to purchase a U.S. vehicle that best meets their governments' requirements in terms of capabilities and cost. Foreign officials said that U.S. manufacturers make vehicles that are reliable and highly capable. When their governments have requirements that align with those associated with U.S. manufactured vehicles, foreign officials said that the U.S. arms transfer control regimes would not be a factor in their governments' decisions to purchase the vehicles. Foreign officials that we interviewed also said their governments are experienced buyers of U.S. arms and are able to successfully navigate the FMS and DCS processes and U.S. end-use restrictions to obtain the military equipment they require.

CONCLUSION

The volume of TWVs DOD purchased to meet operational requirements in Iraq and Afghanistan was unique due to specific threats. Many of these vehicles are no longer needed and DOD's need for new TWVs is expected to decline in coming years. Further, given the current budgetary environment, DOD cannot afford to support the size of its current fleet or buy as many vehicles as it once did. Though U.S. manufacturers increased their production to meet those past needs, they will be challenged in responding to the sharp decline in DOD's TWV requirements in future years. As DOD continues its studies of the U.S. TWV industrial base, it may be better positioned to address these challenges and how DOD can mitigate any risks to sustaining its TWV fleet. It is unlikely that sales to foreign governments will ever offset declines in sales to DOD, but foreign sales may be more important to the industrial base now more than ever. U.S. manufacturers, however, are presented with a number of factors that affect their ability to sell TWVs to foreign governments. While no foreign officials indicated that their governments would not buy U.S. TWVs, there has been relatively limited demand for the vehicles U.S. manufacturers have produced for DOD. Further, there are many foreign manufacturers that can supply vehicles that meet foreign governments' requirements. Each of the U.S. manufacturers we met with was either selling or developing alternative vehicles that better meet foreign governments' requirements, but the extent to which those efforts will stimulate additional sales has yet to be seen. Further, U.S. manufacturers raised concerns that their competitors could eventually include the U.S. military as it makes plans to divest itself of used TWVs that it could make available to foreign governments at reduced costs or for free. Additionally, while U.S. manufacturers perceived the U.S. arms transfer control regimes to be more burdensome than those of other countries, the regimes are not a determining factor when foreign governments seek to purchase TWVs.

AGENCY COMMENTS AND THIRD-PARTY VIEWS

We provided a draft of this report to DOD, State, and Commerce, as well as the four manufacturers and five foreign governments with whom we met, for their review and comment. DOD and State provide technical comments and two of the manufacturers provided clarifications, which we incorporated into the report as appropriate. Commerce, two manufacturers, and the five foreign governments informed us that they had no comments.

Belva M. Martin Director
Acquisition and Sourcing Management

End Notes

[1] For the purposes of this report, the term arms refers to defense articles, defense services, and related technical data as specified by 22 U.S.C. § 2778 and defined in 22 C.F.R. Part 120. Dual-use items are those which have both military and commercial applications.

[2] 22 U.S.C. §§ 2751-2799aa-2.

[3] FMS is limited to eligible countries and international organizations. 22 U.S.C. § 2753.

[4] S. Rep. No. 112–26, at 30 (2011).

[5] We did not include TWV parts purchased by DOD or any efforts intended to sustain or upgrade its existing vehicles.

[6] Licensing data for DCS were obtained from State's DTRADE system, which was implemented in 2008. We did not obtain State licensing data collected prior to its implementation.

[7] 22 U.S.C. § 2778(a).

[8] Exec. Order No. 11,958, 42 Fed. Reg. 4,311 (Jan. 24,1977).

[9] 22 C.F.R. §§ 120 - 130.

[10] 22 C.F.R. § 120.2 and Part 121.

[11] While most defense articles and services on the USML require a license for export, the International Traffic in Arms Regulations contain some exemptions and authorize exceptions.

[12] According to Presidential Decision Directive 34, all arms transfer decisions should take into account the 12 criteria identified in the conventional arms transfer policy.

[13] See, e.g., 22 U.S.C.§ 2753.

[14] Authority granted by the act lapsed on August 20, 2001. 50 U.S.C. app. § 2419. However, Executive Order 13222, Continuation of Export Control Regulations, which was issued in August 2001 under the authority provided by the International Emergency Economic Powers Act (50 U.S.C. § 1702), continues the controls established under the act and the implementing Export Administration Regulations. Executive Order 13222 requires an annual extension and was recently renewed by Presidential Notice on August 15, 2012. 77 Fed. Reg. 49699 (Aug. 16, 2012).

[15] Under the Export Control Reform initiative, State has proposed revisions to Category VII of the USML that, among other things, are intended to more clearly identify ground vehicles controlled on the USML and those controlled on the Commerce Control List. The proposed rule would also move certain spare and replacement parts from the USML to the Commerce Control List. State officials did not have a timeframe for the completion of these proposed changes. 76 Fed. Reg. 76100 (Dec. 6, 2011).

[16] GAO, *Defense Acquisitions: Future Ground-Based Vehicles and Network Initiatives Face Development and Funding Challenges,* GAO-12-181T (Washington, D.C.: Oct. 26, 2011); and *Defense Acquisitions: Issues to Be Considered as DOD Modernizes Its Fleet of Tactical Wheeled Vehicles*, GAO-11-83 (Washington, D.C.: Nov. 5, 2010).

[17] The Army defines reset and recapitalization efforts as actions taken to restore equipment to a desired level of combat capability commensurate with the equipment's future mission. It encompasses maintenance and supply activities that restore and enhance combat capability to equipment that was destroyed, damaged, stressed, or worn out beyond economic repair due to combat operations by repairing, rebuilding, or procuring replacement equipment. Recapitalization rebuilds or repairs equipment to a level that improves the performance capabilities of the equipment or returns the equipment to a "zero mile/zero hour" level with original performance specifications.

[18] This includes the export of TWVs controlled under State or Commerce regulations.

[19] 22 U.S.C. § 2776.

[20] As required by the ACEA, foreign governments may not transfer title to or possession of any defense articles or services to anyone not an officer, employee, or agent of that country unless the country receives prior consent from the U.S. government. 22 U.S.C. § 2753.

INDEX

#

A

B

C

D

E

F

G

H

I

J

K

L

M

N

O

P

Q

R

S